T0315044

Electric Power Systems with Renewables

Simulations Using PSS®E

Electric Power Systems with Renewables

Simulations Using PSS®E

Second Edition

NED MOHAN
University of Minnesota, USA

SWAROOP GUGGILAM
Electric Power Research Institute, USA

Contributors

BRUCE WOLLENBERG

PRATAP MYSORE

DOUGLAS BROWN

Library of Congress Cataloging-in-Publication Data
Names: Mohan, Ned, editor. | Guggilam, Swaroop, editor.
Title: Electric power systems with renewables : simulations using PSS®E /
  edited by Ned Mohan, Swaroop Guggilam.
Description: Second edition. | Hoboken, New Jersey : John Wiley & Sons,
  [2023] | Includes bibliographical references and index.
Identifiers: LCCN 2022045412 (print) | LCCN 2022045413 (ebook) | ISBN
  9781119844877 (hardback) | ISBN 9781119844884 (pdf) | ISBN 9781119844891
  (epub)
Subjects: LCSH: Electric power systems. | Renewable energy sources.
Classification: LCC TK1001 .E26 2023 (print) | LCC TK1001 (ebook) | DDC
  621.31--dc23/eng/20221017
LC record available at https://lccn.loc.gov/2022045412
LC ebook record available at https://lccn.loc.gov/2022045413

Cover Image: © jia yu/Getty Images
Cover Design: Wiley

Set in size of 10/12 and TimesNewRoman MT by Integra Software Services Pvt. Ltd, Pondicherry, India

To our families

CONTENTS

PREFACE   xiii

TABLE OF SIMULATIONS USING PSS®E, PYTHON, AND
         MATLAB/SIMULINK®   xv

ABOUT THE COMPANION WEBSITE   xvii

CHAPTER 1   INTRODUCTION TO POWER SYSTEMS: A
                  CHANGING LANDSCAPE   1

          1.1   Nature of Power Systems   2

          1.2   Changing Landscape of Power Systems Due to
                  Utility Deregulation   4

          1.3   Integration of Renewables Into the Grid   5

          1.4   Topics in Power Systems   6

          References   9

          Problems   9

CHAPTER 2   REVIEW OF BASIC ELECTRIC CIRCUITS AND
                  ELECTROMAGNETIC CONCEPTS   11

          2.1   Introduction   11

          2.2   Phasor Representation in a Sinusoidal Steady
                  State   12

          2.3   Power, Reactive Power, and Power Factor   16

          2.4   Three-Phase Circuits   22

          2.5   Real and Reactive Power Transfer between AC
                  Systems   30

          2.6   Equipment Ratings, Base Values, and
                  Per-Unit Quantities   32

          2.7   Energy Efficiencies of Power System Equipment   33

          2.8   Electromagnetic Concepts   34

	Reference	44
	Problems	44
	Appendix 2A	47

CHAPTER 3	**ELECTRIC ENERGY AND THE ENVIRONMENT**	**51**
	3.1 Introduction	51
	3.2 Choices and Consequences	51
	3.3 Hydropower	53
	3.4 Fossil-Fuel-Based Power Plants	53
	3.5 Nuclear Power	55
	3.6 Renewable Energy	58
	3.7 Distributed Generation (DG)	66
	3.8 Environmental Consequences and Remedial Actions	66
	References	68
	Problems	68

CHAPTER 4	**AC TRANSMISSION LINES AND UNDERGROUND CABLES**	**71**
	4.1 Need for Transmission Lines and Cables	71
	4.2 Overhead AC Transmission Lines	72
	4.3 Transposition of Transmission-Line Phases	73
	4.4 Transmission-Line Parameters	74
	4.5 Distributed-Parameter Representation of Transmission Lines in a Sinusoidal Steady State	82
	4.6 Surge Impedance Z_c and Surge Impedance Loading (SIL)	84
	4.7 Lumped Transmission-Line Models in a Steady State	86
	4.8 Cables	88
	References	89
	Problems	90
	Appendix 4A Long Transmission Lines	92

CHAPTER 5	**POWER FLOW IN POWER SYSTEM NETWORKS**	**95**
	5.1 Introduction	95
	5.2 Description of the Power System	96

5.3	Example Power System	97
5.4	Building the Admittance Matrix	98
5.5	Basic Power-Flow Equations	100
5.6	Newton-Raphson Procedure	101
5.7	Solution of Power-Flow Equations Using the Newton-Raphson Method	104
5.8	Fast Decoupled Newton-Raphson Method for Power Flow	109
5.9	Sensitivity Analysis	110
5.10	Reaching the Bus VAR Limit	110
5.11	Synchronized Phasor Measurements, Phasor Measurement Units (PMUS), and Wide-Area Measurement Systems	111
5.12	DC Power Flow	111
	References	112
	Problems	112
	Appendix 5A Gauss-Seidel Procedure for Power-Flow Calculations	113
	Appendix 5B Remote Bus Voltage Control by Generators	114
CHAPTER 6	**TRANSFORMERS IN POWER SYSTEMS**	**119**
6.1	Introduction	119
6.2	Basic Principles of Transformer Operation	119
6.3	Simplified Transformer Model	125
6.4	Per-Unit Representation	127
6.5	Transformer Efficiencies and Leakage Reactances	131
6.6	Regulation in Transformers	131
6.7	Autotransformers	132
6.8	Phase Shift Introduced by Transformers	134
6.9	Three-Winding Transformers	135
6.10	Three-Phase Transformers	136
6.11	Representing Transformers with Off-Nominal Turns Ratios, Taps, and Phase Shifts	137
6.12	Transformer Model in PSS®E	140
	References	141
	Problems	141

**CHAPTER 7   GRID INTEGRATION OF INVERTER-BASED
                            RESOURCES (IBRS) AND HVDC SYSTEMS   145**

7.1   Climate Crisis   146

7.2   Interface Between Renewables/Batteries and
             The Utility Grid   146

7.3   High-Voltage DC (HVDC) Transmission
             Systems   152

7.4   IEEE P2800 Standard for Interconnection and
             Interoperability of Inverter-Based Resources
             Interconnecting with Associated Transmission
             Electric Power Systems   156

References   157

Problems   157

Appendix 7A   Operation of Voltage Source Converters
                 (VSCS) [7A1]   157

Appendix 7B   Operation of Thyristor-Based Line-
                 Commutated Converters (LCCS)   161

**CHAPTER 8   DISTRIBUTION SYSTEM, LOADS, AND POWER
                            QUALITY   173**

8.1   Introduction   173

8.2   Distribution Systems   173

8.3   Power System Loads   174

8.4   Power Quality Considerations   180

8.5   Load Management   191

References   192

Problems   192

CHAPTER 9   SYNCHRONOUS GENERATORS   195

9.1   Introduction   195

9.2   Structure   196

9.3   Induced EMF in the Stator Windings   200

9.4   Power Output, Stability, and The Loss of
             Synchronism   204

9.5   Field Excitation Control to Adjust
             Reactive Power   206

9.6   Field Exciters for Automatic Voltage
             Regulation (AVR)   208

9.7 Synchronous, Transient, and Subtransient
Reactances 208

9.8 Generator Modeling in PSS®E 211

References 213

Problems 213

CHAPTER 10 VOLTAGE REGULATION AND STABILITY IN
POWER SYSTEMS 215

10.1 Introduction 215

10.2 Radial System as an Example 215

10.3 Voltage Collapse 218

10.4 Preventing Voltage Instability 220

References 227

Problems 228

CHAPTER 11 TRANSIENT AND DYNAMIC STABILITY
OF POWER SYSTEMS 229

11.1 Introduction 229

11.2 Principle of Transient Stability 229

11.3 Transient Stability Evaluation
in Large Systems 238

11.4 Dynamic Stability 239

References 240

Problems 241

Appendix 11A Inertia, Torque, and Acceleration in
Rotating Systems 241

CHAPTER 12 CONTROL OF INTERCONNECTED POWER
SYSTEMS AND ECONOMIC DISPATCH 245

12.1 Control Objectives 245

12.2 Voltage Control by Controlling Excitation and
Reactive Power 246

12.3 Automatic Generation Control (AGC) 247

12.4 Economic Dispatch and Optimum Power Flow 257

References 262

Problems 262

CHAPTER 13 TRANSMISSION LINE FAULTS, RELAYING, AND CIRCUIT BREAKERS **265**

13.1 Causes of Transmission Line Faults 265

13.2 Symmetrical Components for Fault Analysis 266

13.3 Types of Faults 269

13.4 System Impedances for Fault Calculations 273

13.5 Calculating Fault Currents in Large Networks 276

13.6 Protection Against Short-Circuit Faults 277

References 286

Problems 287

CHAPTER 14 TRANSIENT OVERVOLTAGES, SURGE PROTECTION, AND INSULATION COORDINATION **289**

14.1 Introduction 289

14.2 Causes of Overvoltages 289

14.3 Transmission-Line Characteristics and Representation 292

14.4 Insulation to Withstand Overvoltages 294

14.5 Surge Arresters and Insulation Coordination 296

References 296

Problems 297

INDEX **299**

PREFACE

Role of Electric Power Systems in Sustainability

It is estimated that approximately 40% of the energy used in the United States is first converted into electricity. This percentage will grow to 60–70% if we begin to use electricity for transportation by means of high-speed trains and electric and electric-hybrid vehicles. Of course, generating electricity by using renewables and using it efficiently are extremely important for sustainability. In addition, electricity is often generated in areas far from where it is used, and therefore how efficiently and reliably it is delivered is equally important for sustainability.

Lately there has been a great deal of emphasis on the smart grid, whose definition remains somewhat vague. Nonetheless, we can all agree that we need to allow the integration of electricity harnessed from renewables, such as solar and wind, and storage into the grid and deliver it reliably and efficiently. To derive the benefits of such possibilities, a thorough understanding of how electric power networks operate is extremely important, and that is the purpose of this textbook.

The subject of electric power systems encompasses a large and complex set of topics. An important aspect of this textbook is a balanced approach in presenting as many topics as deemed relevant on a fundamental basis for a single-semester course. These topics include how electricity is generated, how it is used by various loads, and the network and equipment in between. Students will see the big picture and simultaneously learn the fundamentals. The topic sequence has been carefully considered to avoid repetition and retain students' interest. However, instructors can rearrange the order based on their own experience and preference.

In a fast-paced course like this, student learning can be significantly enhanced by computer simulations. We have used PSS®E, a simulation software widely used in many countries. However, the knowledge and concepts apply to any other power-system simulation software.

The authors are indebted to Professor Bruce Wollenberg, Pratap Mysore, and Douglas Brown for their valuable contributions and Dr. Madhukar Rao Airineni for his help during the preparation of this book.

TABLE OF SIMULATIONS USING PSS®E, PYTHON, AND MATLAB/SIMULINK®

Topic	Simulation	Page
Transmission-line constants	Example 4.2 using PSS®E	81
Power flow in a three-bus system	Example 5.4 using PSS®E, Python, and MATLAB	107
DC power flow	Example 5.5 using Python and MATLAB	112
N-R power flow	Example 5.6 using PSS®E, Python, and MATLAB	116
Modeling a transformer in power flow	Example 6.1 using PSS®E	129
Modeling IBRs	Example 7.1 using PSS®E	151
Modeling an HVDC line	Example 7.2 using PSS®E	154
Modeling generators in power flow	Chapter 9 using PSS®E	211
Modeling a STATCOM in power flow	Example 10.1 using PSS®E	225
Transient stability	Example 11.1 using PSS®E, Python, and MATLAB	233
Interconnected systems	Example 12.3 using Simulink®	255
Optimal power flow	Chapter 12 using PSS®E	261
Fault calculations	Example 13.1, 13.2, 13.3 using Python and MATLAB	268, 275, 277

Bonus chapter on Python and PSS®E Python scripting: See the accompanying website.

ABOUT THE COMPANION WEBSITE

This book is accompanied by a companion website which includes a number of resources created by author for students and instructors that you will find helpful.

wiley.com/go/mohaneps

The Instructor website includes the following resources for each chapter:

- Videos
- Solution Manual for the Problems at the end of each Chapter.
- Slides for Lectures in PDF form.

The Student website includes the following resources for each chapter:

- Videos
- Bonus PDF Chapter on "Using Python with PSS®E"
- Live Web Page Link (To showcase day-to-day activities happening around the book, like workshops, live sessions, etc.)

Please note that the resources in instructor website are password protected and can only be accessed by instructors who register with the site.

1

INTRODUCTION TO POWER SYSTEMS: A CHANGING LANDSCAPE

Electric power systems are technical wonders; and according to the National Academy of Engineering [1], electricity and its accessibility are the greatest engineering achievements of the twentieth century, ahead of computers and airplanes. In many respects, electricity is a basic human right. It is a highly refined "commodity," without which it is difficult to imagine how a modern society could function. It has saved countless millions from the daily drudgery of backbreaking menial tasks.

Unfortunately, a billion people in the world have either no access or no reliable access to electricity [2]. Added to this challenge is the fact that burning fossil fuels such as coal and natural gas to produce electricity results in carbon dioxide and other greenhouse gases. These greenhouse gases are causing global warming and climate change, the gravest threat facing human civilization.

Therefore, we as electric power engineers are faced with twin challenges. How we generate electricity using renewables such as wind and solar, how we transmit and deliver it, and how we use it are key factors to meet these challenges.

Electric Power Systems with Renewables: Simulations Using PSS®E, Second Edition. Ned Mohan and Swaroop Guggilam.
© 2023 John Wiley & Sons, Inc. Published 2023 by John Wiley & Sons, Inc.
Companion Website: www.wiley.com/go/mohaneps

1.1   NATURE OF POWER SYSTEMS

Power systems encompass the generation of electricity to its ultimate consumption in operating everything from computers to hairdryers. In the most simplistic form, a power system is shown in Figure 1.1, where power from a single generating station is being supplied to consumers.

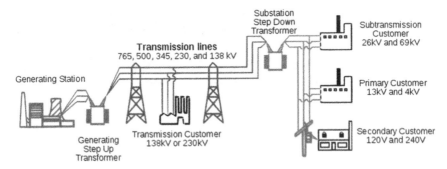

FIGURE 1.1   A single generating station supplying consumers (in color on the accompanying website). Source: [3] / U.S Department of Energy / Public Domain.

The system shown in Figure 1.1 is for illustration purposes only and shows the various components of a power system if such a system were to be constructed. It consists of a generating station, possibly producing voltages at a 20 kV level, a transformer that steps up this voltage to much higher transmission voltages for long-distance transmission of power, and then another transformer to step down the voltage to supply consumers at various voltages. In this book, we will look at all these components.

However, as mentioned, the system in Figure 1.1 is for illustration only. In practice, for example, the North American grid in the United States and Canada consists of thousands of generators, all operating in synchronism. These generators are interconnected by over 200,000 miles of transmission lines at 230 kV voltage levels and above, as shown in Figure 1.2. Such an interconnected system results in the continuity and reliability of service if there is an outage in one part of the system and provides electricity at the lowest cost by utilizing the lowest-cost generation as much as possible at a given time.

This power system has evolved over several decades, and a good history of it can be read in [5].

As mentioned earlier, even though the actual power system may consist of tens of thousands of generators and hundreds of thousands of miles of transmission lines, it is possible to zoom in on a subset of such an extremely large system. This is illustrated in Figure 1.3, as an example, which consists of only 10 generators. Although power transmission systems are always three-phase

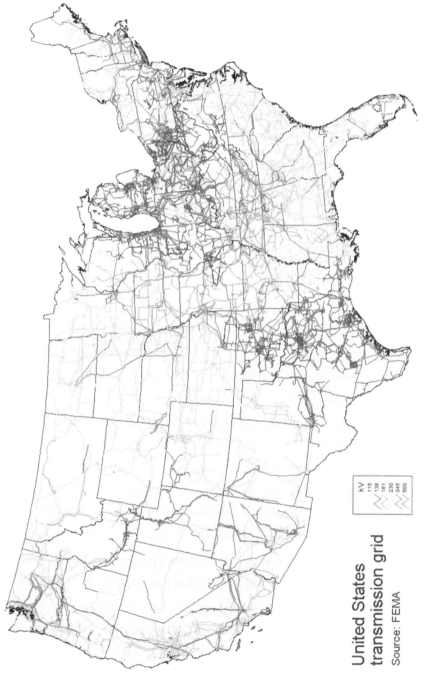

United States
transmission grid

Source: FEMA

kV
115
138
161
230
345
500

FIGURE 1.2 Interconnected North American power grid (in color on the accompanying website). Source: [4].

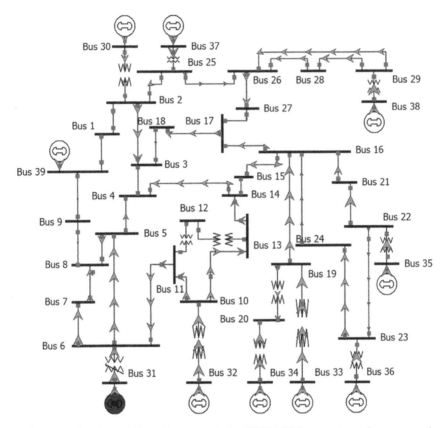

FIGURE 1.3 A one-line diagram of the IEEE 39 bus system, known as the 10-machine New England Power System (in color on the accompanying website). It has 10 generators and 46 lines. Source: [6].

(except in high-voltage DC [HVDC] transmission systems), we represent them with one line in the figure, in a so-called *one-line diagram*.

1.2 CHANGING LANDSCAPE OF POWER SYSTEMS DUE TO UTILITY DEREGULATION

Power systems today are undergoing major changes in how they are evolving in their structure and meeting load demand. In the past (and still true to some extent), electric utilities were highly centralized, owning large central power plants as well as the transmission and distribution systems, all the way down to the consumer loads. These utilities were monopolies: consumers had no choice but to buy power from their local utilities. For oversight purposes, utilities were highly regulated by Public Service Commissions that acted as consumer watchdogs, preventing utilities from price gouging, and as custodians of the environment by not allowing avoidable polluting practices.

The structure and operation of power systems are beginning to change, and the utilities have been divided into separate generation and transmission/distribution companies. There is distributed generation (DG) by independent power producers (IPPs), and there are distributed energy resources (DERs) to generate electricity by whatever means (wind, for example); they must be allowed access to the transmission grid to sell power to consumers. The impetus for the breakup of the utility structure was provided by the enormous benefits of deregulation in the telecommunication and airline industries, which fostered a large degree of competition, resulting in much lower rates and much better service to consumers. Despite the inherent differences between these two industries and the utility industry, it was perceived that utility deregulation would similarly profit consumers with lower electricity rates.

This deregulation is in transition, with some states and countries pursuing it more aggressively and others more cautiously. To promote open competition, utilities are forced to restructure by unbundling their generation units from their transmission and distribution units. The objective is that the independent transmission system operators (TSOs) wheel power for a charge from anywhere and from anyone to the customer site. This fosters competition, allowing open transmission access to everyone: for example, IPPs. Many such small IPPs have gone into business, producing power using gas turbines, windmills, and PV plants.

Operation in a reliable manner is ensured by independent system operators (ISOs), and financial transactions are governed by real-time bidding to buy and sell power. Energy traders have gotten into the act for profit: buying energy at lower prices and selling it at higher prices in the spot market. Utilities are signing long-term contracts for energy, such as gas. This is all based on the rules of the financial world: forecasting, risks, options, reliability, etc.

As mentioned earlier, the outcome of this deregulation, still in transition, is far from certain. However, there is every reason to believe that the deregulation now in progress will continue, with little possibility that the clock will be turned back. Some fixes are needed. The transmission grid has become a bottleneck, with little financial incentive for TSOs to increase capacity. If the transmission system is congested, TSOs can charge higher prices. The number of transactions and the complexity of these transactions have increased dramatically. These factors point to anticipated legislative actions needed to maintain electric system reliability.

1.3 INTEGRATION OF RENEWABLES INTO THE GRID

In addition to the deregulation mentioned, there is a great deal of emphasis on generating power using renewables such as wind and solar rather than fossil fuels such as coal and natural gas that emit greenhouse gases. The cost of power from these renewables has been declining and, in many cases, is lower than the cost of conventional sources. In making this comparison, we must realize that renewables are intermittent, and thus their value goes down as their penetration into the grid increases.

At present, the amount of electricity produced by renewables is small, as shown in Figure 1.4 for the United States.

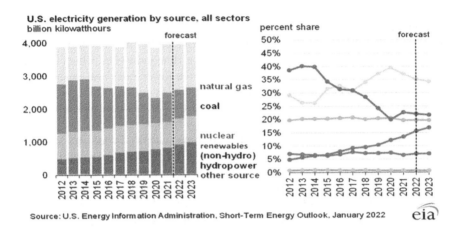

Source: U.S. Energy Information Administration, Short-Term Energy Outlook, January 2022

FIGURE 1.4   Generation of electricity by various sources in the United States (in color on the accompanying website). Source: [7].

However, due to climate concerns, the portion of electricity from renewable sources will undoubtedly grow, and our study of power systems must include how we can accommodate them in the grid.

1.4   TOPICS IN POWER SYSTEMS

The purpose of this textbook is to provide a complete overview of power systems meeting present and future energy needs. As we can appreciate, the interconnected power system with thousands of generators and hundreds of thousands of transmission lines between them is vast and complex. Therefore, the question in front of us is how we can impart the fundamental concepts and learn the workings of various components while pointing to the real tools used in industry to study such systems in their entirety.

It should be recognized that there can be planning studies that may have over 90,000 buses—e.g. the entire Eastern Interconnection System in the United States. However, the authors have taken the three-bus example shown in Figure 1.5 to explore various fundamental concepts. To extend these concepts to the study of the real system, the authors have decided to use PSS®E [8] from Siemens, which is one of the most widely used software packages in the utility industry in over 140 countries. The analysis of this three-bus simple system is shown in Figure 1.6 using PSS®E.

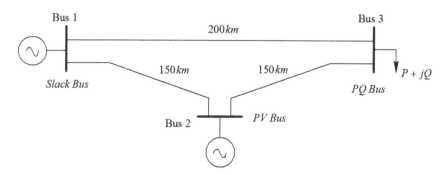

FIGURE 1.5 A three-bus example system.

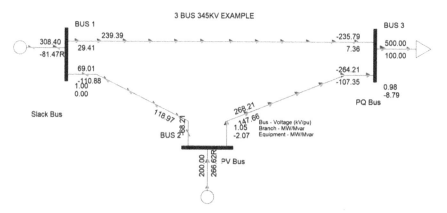

FIGURE 1.6 Simulation of the three-bus example system in Figure 1.5 using PSS®E.

The topics covered by this book's chapters are described next and arranged to associate the lecture material with the laboratory exercises chronologically. All of these topics are supplemented by simulations in PSS®E as appropriate.

Chapter 2: This chapter describes the basic concepts that are fundamental to the analysis of power system circuits. These include phasor representation in sinusoidal steady state, power, reactive power, and power factors. The chapter describes the three-phase circuit analysis, expressing quantities in per-unit, the energy efficiency of power system apparatus, and electromagnetic concepts essential to understanding transformers and electrical generators.

Chapter 3: We have a choice of using various resources for generating electricity, but there are always consequences to any selection. These are briefly discussed, including hydro plants, fossil fuel–based power plants, nuclear power, and the increasing role of renewable energy sources such as wind and solar. The environmental consequences of these choices are discussed.

Chapter 4: This chapter describes the need for transmission lines and cables, AC transmission lines and their parameters, and various representations. It also includes a brief discussion of cables. It shows the use of PSS®E for calculating line constants.

Chapter 5: For the purposes of planning and operating securely under contingencies caused by outages, it is important to know how power flows on various transmission lines. This chapter describes various power-flow techniques that include the Newton-Raphson method, the fast-decoupled technique, the Gauss-Siedel approach, and the DC power-flow method. Examples using PSS®E are given.

Chapter 6: Voltages produced by generators are stepped up by transformers for long-distance hauling of power over transmission lines. This chapter includes basic principles of transformer operation, simplified transformer models, per-unit representation, regulation, phase shifts introduced by transformers, and auto-transformers. It shows how transformers are represented in calculating power flow using PSS®E.

Chapter 7: This chapter describes the role of inverter-based resources (IBRs) and HVDC transmission systems, including those using voltage source converters (VSCs) and line-commutated converters (LCCs). It also includes a brief discussion of the IEEE P2800. Examples using PSS®E are given for integrating IBRs and using HVDC-VSC.

Chapter 8: This chapter describes consumer loads and the role of power electronics, which are changing their nature. It also describes how these loads react to voltage fluctuations and their impact on power quality.

Chapter 9: To generate electricity, steam and natural gas are utilized to run turbines that provide mechanical input to synchronous generators to produce three-phase electrical voltages. Synchronous generators are described in this chapter. It shows how generators are represented for power flow and transient stability analysis.

Chapter 10: Transmission lines are being loaded more than ever, making voltage stability a concern, as discussed in this chapter. Power electronics have a growing role in power systems in the form of flexible AC transmission systems (FACTS), which are described in this chapter for improving voltage stability. It includes an example of adding a static synchronous compensator (STATCOM) at a bus in the power-flow analysis using PSS®E.

Chapter 11: Maintaining stability so that various generators operate in synchronism is described in this chapter, which discusses how the stability in an interconnected system, with thousands of generators operating in synchronism, can be maintained in response to transient conditions, such as transmission-line faults, when there is a mismatch between the mechanical power input to the turbines and the electrical power that can be transmitted.

Chapter 12: This chapter discusses economic dispatch, where generators are loaded in such a way as to provide overall economy of operation. The operation of interconnected systems is also described, so that the power system frequency and voltages are maintained at their nominal values and purchasing and selling agreements between various utilities are honored. An example of optimal power flow (OPF) is given using PSS®E.

Chapter 13: Power systems are spread over large areas. Being exposed to the elements of nature, they are subjected to occasional faults against which they must be designed and protected so that such events result only in momentary loss of power and no permanent equipment damage accrues. Short circuits on transmission systems are discussed in this chapter, which describes how relays detect faults and cause circuit breakers to open the circuit, interrupting the fault current and then reclosing the circuit breakers to bring the operation back to normal as soon as possible.

Chapter 14: Lightning strikes and switching of extra-high-voltage transmission lines during reenergizing, particularly with trapped charge, can result in very high-voltage surges, which can cause insulation to flash over. To avoid this, surge arresters are used and are properly coordinated with the insulation level of the power systems apparatus to prevent damage. These topics are discussed in this chapter.

REFERENCES

1. National Academy of Engineering. www.nae.edu.
2. United Nations Energy. https://un-energy.org/.
3. US Department of Energy. 2003. Final report, "Blackout in the United States and Canada.".
4. Endeavor MAPSearch. www.mapsearch.com.
5. Julie Cohn. 2017. *The Grid*. The MIT Press.
6. IEEE 39-Bus System. https://electricgrids.engr.tamu.edu/electric-grid-test-cases/ieee-39-bus-system. Provided by Texas A&M University researchers free for commercial or non-commercial use.
7. US Energy Information Administration (EIA). www.eia.gov.
8. Siemens Global. PSS®E high-performance transmission planning and analysis software.https://new.siemens.com/global/en/products/energy/energy-automation-and-smart-grid/pss-software/pss-e.html.

PROBLEMS

1.1   What are the advantages of a highly interconnected system?
1.2   What are the changes taking place in the utility industry?
1.3   What is meant by the following terms: DG, DER, IPP, TSO, and ISO?
1.4   What are the different topics in power systems for understanding its basic nature?

2

REVIEW OF BASIC ELECTRIC CIRCUITS AND ELECTROMAGNETIC CONCEPTS

2.1 INTRODUCTION

The purpose of this chapter is to review elements of basic electric circuit theory that are essential to the study of electric power circuits: using phasors to analyze circuits in a sinusoidal steady state, real and reactive powers, the power factor, analysis of three-phase circuits, power flow in AC circuits, and per-unit quantities [1].

In this book, we use MKS units and IEEE-standard letters and graphic symbols whenever possible. The lowercase letters v and i are used to represent instantaneous values of voltages and currents that vary as functions of time. A current's positive direction is indicated by an arrow, as shown in Figure 2.1. Similarly, the voltage polarities must be indicated. The voltage v_{ab} refers to the voltage of node a with respect to node b, and thus $v_{ab} = v_a - v_b$.

Electric Power Systems with Renewables: Simulations Using PSS®E, Second Edition. Ned Mohan and Swaroop Guggilam.

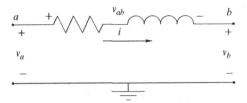

FIGURE 2.1   Convention for voltages and currents.

2.2   PHASOR REPRESENTATION IN A SINUSOIDAL STEADY STATE

In linear circuits with sinusoidal voltages and currents of frequency f applied long enough that a steady state has been reached, all circuit voltages and currents are at a frequency $f(=\omega/2\pi)$. To analyze such circuits, the calculations are simplified through phasor-domain analysis. Using phasors also provides a deeper insight into circuit behavior relatively easily.

In the phasor domain, the time-domain variables $v(t)$ and $i(t)$ are transformed into phasors represented by the complex variables $\bar{V}$ and $\bar{I}$. Note that these phasors are expressed using uppercase letters with a bar (–) on top. In a complex (real and imaginary) plane, these phasors can be drawn with a magnitude and an angle. A co-sinusoidal time function is taken as a reference phasor that is entirely real with an angle of zero degrees. Therefore, the time-domain voltage expression in Equation 2.1 is represented by a corresponding phasor

$$v(t) = \sqrt{2}\,V\cos(\omega t + \phi_v) \rightarrow \Leftrightarrow \rightarrow \bar{V} = V\angle\phi_v \tag{2.1}$$

Similarly,

$$i(t) = \sqrt{2}\,I\cos(\omega t + \phi_i) \rightarrow \Leftrightarrow \rightarrow \bar{I} = I\angle\phi_i \tag{2.2}$$

where V and I are the rms values of the voltage and current. These voltage and current phasors are drawn in Figure 2.2. In Equations 2.1 and 2.2, the angular frequency ω is implicitly associated with each phasor. Knowing this frequency, a phasor expression can be re-transformed into a time-domain expression.

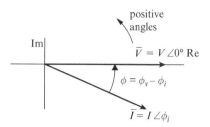

FIGURE 2.2   Phasor diagram. Here, $\phi_v = 0°$; ϕ_i = negative; $\phi = (\phi_v - \phi_i) = -\phi_i$ = positive.

Using phasors, we can convert differential equations into easily solvable algebraic equations containing complex variables. Consider the circuit in Figure 2.3a in a sinusoidal steady state with an applied voltage at a frequency $f(= \omega / 2\pi)$.

To calculate the current in this circuit, remaining in the time domain would require solving the following differential equation:

$$Ri(t) + L\frac{di(t)}{dt} + \frac{1}{C}\int i(t) \cdot dt = \sqrt{2}\,V\cos(\omega t) \tag{2.3}$$

Using phasors, we can redraw the circuit from Figure 2.3a as Figure 2.3b, where the inductance L is represented by its reactance $X_L = \omega L$ and its impedance $Z_L = jX_L$. Similarly, the capacitance C is represented by its reactance $X_C = \left(-\dfrac{1}{\omega C}\right)$ and its impedance $Z_C = jX_C$.

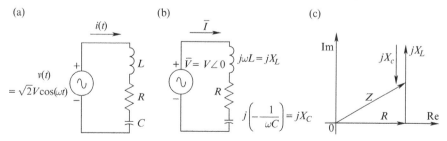

FIGURE 2.3   A circuit (a) in the time domain and (b) in the phasor domain; (c) impedance triangle.

In the phasor-domain circuit, the impedance Z of the series-connected elements is obtained by the impedance triangle in Figures 2.3c as

$$Z = R + \underbrace{jX_L}_{Z_L} + \underbrace{jX_c}_{Z_c} \tag{2.4}$$

where the reactance X is the imaginary part of an impedance Z and therefore,

$$X_L = \omega L \text{ and } X_c = \left(-\frac{1}{\omega C}\right) \tag{2.5}$$

This impedance can be expressed as

$$Z = |Z| \angle \phi \tag{2.6a}$$

where

$$|Z| = \sqrt{R^2 + \left(\underbrace{\omega L}_{X_L} + \underbrace{\left(-\frac{1}{\omega C}\right)}_{X_c}\right)^2} \rightarrow \text{and } \phi = \tan^{-1}\left[\frac{\left(\omega L + \left(-\frac{1}{\omega C}\right)\right)}{R}\right] \tag{2.6b}$$

It is important to recognize that while Z is a complex quantity, it is *not* a phasor and therefore does *not* have a corresponding time-domain expression.

Example 2.1
Calculate the impedance seen from the terminals of the circuit in Figure 2.4 under a sinusoidal steady state at a frequency $f = 60\text{Hz}$.

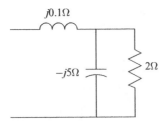

FIGURE 2.4   Impedance network for Example 2.1.

Solution $Z = j0.1 + \dfrac{2 \times (-j5)}{(2 - j5)} = 1.72 - j0.59 = 1.82\angle - 18.9° \ \Omega$.

Using the impedance in Equation 2.6, and assuming that the voltage phase angle ϕ_v is zero, the current in Figure 2.3b can be obtained as

$$\bar{I} = \frac{\bar{V}}{Z} = \left(\frac{V}{|Z|}\right)\angle -\phi \tag{2.7}$$

where $I = \dfrac{V}{|Z|}$ and ϕ is as calculated from Equation 2.6b. Using Equation 2.2, the current can be expressed in the time domain as

$$i(t) = \frac{\sqrt{2}\,V}{|Z|}\cos(\omega t - \phi) \tag{2.8}$$

In the impedance triangle in Figure 2.3c, a positive value of the phase angle ϕ implies that the current lags the voltage in the circuit in Figure 2.3a. Sometimes it is convenient to express the inverse of the impedance, which is called *admittance*:

$$Y = \frac{1}{Z} \tag{2.9}$$

The phasor-domain procedure for solving $i(t)$ is much easier than solving the differential-integral equation given by Equation 2.3.

Example 2.2
Calculate the current $\bar{I}_1$ and $i_1(t)$ in the circuit in Figure 2.5 if the applied voltage has an rms value of 120 V and a frequency of 60 Hz. Assume $\bar{V}_1$ to be the reference phasor.

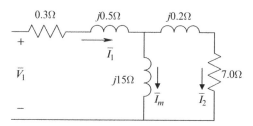

FIGURE 2.5   Circuit for Example 2.2.

Solution With $\bar{V}_1$ as the reference phasor, it can be written as $\bar{V}_1 = 120\angle 0°\text{V}$. The input impedance of the circuit seen from the applied voltage terminals is

$$Z_{in} = (0.3 + j0.5) + \frac{(j15)(7 + j0.2)}{(j15) + (7 + j0.2)} = 6.775\angle 29.03°\ \Omega$$

Therefore, the current $\overline{I}_1$ can be obtained as

$$\overline{I}_1 = \frac{\overline{V}_1}{Z_{in}} = \frac{120\angle 0°}{6.775\angle 29.03°} = 17.712\angle -29.03° \, \mathrm{A}$$

and hence $i_1(t) = 25.048\cos(\omega t - 29.03°)\,\mathrm{A}$.

2.3 POWER, REACTIVE POWER, AND POWER FACTOR

Consider the generic circuit in Figure 2.6 in a sinusoidal steady state. Each sub-circuit may consist of passive (*R-L-C*) elements and active voltage and current sources. Based on the arbitrarily chosen voltage polarity and the current direction shown in Figure 2.6, the instantaneous power $p(t) = v(t)i(t)$ is delivered by subcircuit 1 and absorbed by subcircuit 2. This is because in subcircuit 1, the positively defined current comes out of the positive-polarity terminal (the same as in a generator). On the other hand, the positively defined current enters the positive-polarity terminal in subcircuit 2 (the same as in a load). A negative value of $p(t)$ reverses the roles of subcircuits 1 and 2.

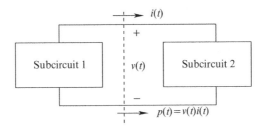

FIGURE 2.6   A generic circuit divided into two subcircuits.

Under a sinusoidal steady state condition at a frequency f, the complex power S, the reactive power Q, and the power factor express how effectively the real (average) power P is transferred from one subcircuit to the other. If $v(t)$ and $i(t)$ are in phase, the instantaneous power $p(t) = v(t)i(t)$ shown in Figure 2.7a pulsates at twice the steady-state frequency, as shown here (V and I are the rms values)

$$\begin{aligned} p(t) &= \sqrt{2}V\cos\omega t \cdot \sqrt{2}I\cos\omega t = 2VI\cos^2\omega t \\ &= VI + VI\cos 2\omega t \,(i \text{ in phase with } v) \end{aligned} \tag{2.10}$$

where both ϕ_v and ϕ_i are assumed to be zero without any loss of generality. In this case, $p(t) \geq 0$ at all times, and therefore the power always flows in one direction: from subcircuit 1 to subcircuit 2. The average over one cycle of the second term on the right side of Equation 2.10 is zero; therefore, the average power is $P = VI$.

Now consider the waveforms shown in Figure 2.7b, where the $i(t)$ waveform lags behind the $v(t)$ waveform by a phase angle ϕ $(= \phi_v - \phi_i)$. Here, $p(t)$ becomes negative during a time interval of (ϕ / ω) during each half-cycle, as calculated here:

$$p(t) = \sqrt{2}V \cos \omega t \cdot \sqrt{2}I \cos(\omega t - \phi) = VI \cos \phi + VI \cos(2\omega t - \phi) \quad (2.11)$$

A negative instantaneous power implies power flow in the opposite direction. This back-and-forth flow of power indicates that the real (average) power is not optimally transferred from one subcircuit to the other, as in Figure 2.7a. Therefore, the average power $P(= VI \cos \phi)$ in Figure 2.7b is less than that in Figure 2.7a, even though the peak voltage and current values are the same in both situations.

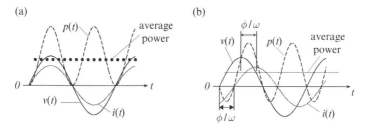

FIGURE 2.7 Instantaneous power with sinusoidal currents and voltages. (a) Voltage and current are in phase; (b) current lags behind voltage.

The circuit from Figure 2.6 is redrawn in Figure 2.8a in the phasor domain. The voltage and the current phasors are defined by their magnitudes and phase angles as

$$\bar{V} = V \angle \phi_v \rightarrow \text{ and } \rightarrow \bar{I} = I \angle \phi_i \quad (2.12)$$

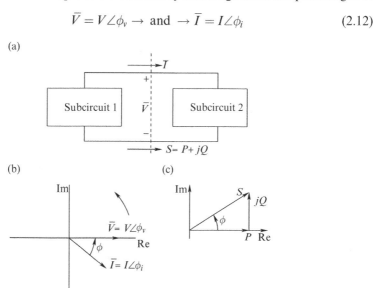

FIGURE 2.8 (a) Circuit in the phasor domain; (b) phasor diagram with $\phi_v = 0$; (c) power triangle.

The complex power S is defined as

$$S = \bar{V}\bar{I}^* \rightarrow (* \text{ indicates complex conjugate}) \qquad (2.13)$$

Therefore, substituting the expressions for voltage and current into Equation 2.13, and noting that $\bar{I}^* = I\angle - \phi_i$,

$$S = V\angle\phi_v\, I\angle - \phi_i = VI\angle(\phi_v - \phi_i) \qquad (2.14)$$

The difference between the two phase angles is defined as before

$$\phi = \phi_v - \phi_i \qquad (2.15)$$

Therefore,

$$S = VI\angle\phi = P + jQ \qquad (2.16)$$

where

$$P = VI\cos\phi \qquad (2.17)$$

$$Q = VI\sin\phi \qquad (2.18)$$

In Equation 2.17, $I\cos\phi$ is the current component that is in phase with the voltage phasor in Figure 2.8b and results in real power transfer P. On the other hand, from Equation 2.18, $I\sin\phi$ is the current component that is at an angle of 90 degrees to the voltage phasor in Figure 2.8b and results in a reactive power Q but no average real power.

The power triangle corresponding to the phasors in Figure 2.8b is shown in Figure 2.8c. From Equation 2.16, the magnitude of S, also called the *apparent power*, is

$$|S| = \sqrt{P^2 + Q^2} \qquad (2.19)$$

and

$$\phi = \tan^{-1}\left(\frac{Q}{P}\right) \qquad (2.20)$$

These quantities have the following units: P, W (watts); Q, var (volt-amperes reactive); $|S|$, VA (volt-amperes); and ϕ_v, ϕ_i, ϕ, radians, measured positively in a counterclockwise direction with respect to the real axis that is drawn horizontally from left to right.

The physical significance of the apparent power $|S|$, P, and Q should be understood. The cost of most electrical equipment such as generators, transformers, and transmission lines is proportional to $|S|(= VI)$ since their electrical insulation level and the magnetic core size for a given line frequency depend on the voltage V, and the conductor size depends on the rms current I. The real power P has a physical significance since it represents the useful work being performed plus the losses. Under most operating conditions, it is desirable to have the reactive power Q be zero; otherwise, it results in increased $|S|$.

To support this discussion, another quantity called the *power factor* is defined. The power factor is a measure of how effectively a load draws real power:

$$\text{power factor} = \frac{P}{|S|} = \frac{P}{VI} = \cos\phi \tag{2.21}$$

This is a dimension-less quantity. Ideally, the power factor should be 1.0 (that is, Q should be zero) to draw real power with a minimum current magnitude and hence minimize losses in electrical equipment such as generators, transformers, and transmission and distribution lines. An inductive load draws power at a lagging power factor where the load current lags behind the voltage. Conversely, a capacitive load draws power at a leading power factor where the load current leads the load voltage.

Example 2.3

Calculate P, Q, S, and the power factor of operation at the terminals in the circuit shown in Figure 2.5 in Example 2.2.

Solution

$$P = V_1 I_1 \cos\phi = 120 \times 17.712 \cos 29.03° = 1858.4 \text{ W}$$

$$Q = V_1 I_1 \sin\phi = 120 \times 17.712 \times \sin 29.03° = 1031.3 \text{ var}$$

$$|S| = V_1 I_1 = 120 \times 17.7 = 2125.4 \text{ VA}$$

From Equation 2.20, $\phi = \tan^{-1}\dfrac{Q}{P} = 29.03°$ in the power triangle shown in Figure 2.8c. Note that the angle of S in the power triangle is the same as the impedance angle ϕ in Example 2.2. The power factor of operation is

$$\text{power factor} = \cos\phi = 0.874$$

Note the following for the inductive impedance in Example 2.3: (i) the impedance is $Z = |Z| \angle\phi$, where ϕ is positive. (ii) The current lags the voltage by the impedance angle ϕ. This corresponds to a lagging power factor of operation. (iii) In the power triangle, the impedance angle ϕ relates P, Q, and S. (iv) An inductive impedance, when applied to a voltage source, draws a positive reactive power Q_L (var). If the impedance were to be capacitive, the phase angle ϕ

would be negative, and this impedance, when a voltage was applied, would draw a negative reactive power Q_C (that is, this impedance would *supply* a positive reactive power).

2.3.1 Sum of Real and Reactive Powers in a Circuit

In a circuit, all the real powers supplied to various components sum to the total real power supplied:

$$\text{total real power supplied} = \sum_k P_k = \sum_k I_k^2 R_k \qquad (2.22)$$

Similarly, all the reactive powers supplied to various components sum to the total reactive power supplied

$$\text{total reactive power supplied} = \sum_k Q_k = \sum_k I_k^2 X_k \qquad (2.23)$$

where X_k and hence Q_k is negative if a component is capacitive.

Example 2.4
In the circuit shown in Figure 2.5 in Example 2.2, calculate P and Q associated with each element and calculate the total real and reactive powers supplied at the terminals. Confirm these calculations by comparing with P and Q calculated in Example 2.3.

Solution From Example 2.2, $\bar{I}_1 = 17.712\angle -29.03°\,\text{A}$.

$$\bar{I}_m = \bar{I}_1 \frac{R_2 + jX_2}{(R_2 + jX_2) + jX_m} = 7.412\angle -92.66°\,\text{A, and } \bar{I}_2 = \bar{I}_1 - \bar{I}_m = 15.876\angle -4.3°\,\text{A}.$$

Therefore, $P_{R_1} = R_1 I_1^2 = 0.3 \times 17.172^2 = 94.11\,\text{W}$, $P_{R_2} = R_2 I_2^2 = 7 \times 15.876^2 = 1764.3\,\text{W}$, and

$$\sum P = P_{R_1} + P_{R_2} = 1858.4\,\text{W}$$

For the reactive vars, $Q_{X_1} = X_1 I_1^2 = 0.5 \times 17.172^2 = 156.851\,\text{var}$,

$$Q_{X_2} = X_2 I_2^2 = 0.2 \times 15.876^2 = 50.409\,\text{var and}$$
$$Q_{X_3} = X_m I_m^2 = 15 \times 7.412^2 = 821.021\,\text{var}$$

Therefore,

$$\sum Q = Q_{X_1} + Q_{X_2} + Q_{X_m} = 1031.3\,\text{var}$$

Note that $\sum P$ and $\sum Q$ are equal to P and Q at the terminals, as calculated in Example 2.3.

2.3.2 Power Factor Correction

As explained earlier, utilities prefer that loads draw power at the unity power factor so the current for a given power drawn is minimum, thus resulting in the least amount of I^2R losses in the resistances associated with transmission and distribution lines and other equipment. This power-factor correction can be accomplished by compensating or nullifying the reactive power drawn by the load by connecting a reactance in parallel that draws the same reactive power in magnitude but of the opposite sign. This is illustrated by the following example.

Example 2.5
In the circuit shown in Figure 2.5 in Example 2.3, the complex power drawn by the load impedance was calculated as $P_L + jQ_L = (1858.4 + j1031.3)\,\text{VA}$. Calculate the capacitive reactance in parallel that will result in the overall power factor being unity, as seen from the voltage source.

Solution The load is drawing reactive power $Q_L = 1031.3\,\text{var}$. Therefore, as shown in Figure 2.9, a capacitive reactance must be connected in parallel with the load impedance to draw $Q_C = -1031.3\,\text{var}$ (or *supply* positive reactive var equal to Q_L) such that only the real power is drawn from the voltage source.

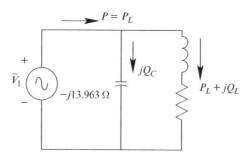

FIGURE 2.9 Power factor correction for Example 2.5.

Since the voltage across the capacitive reactance is given, the capacitive reactance can be calculated as

$$X_C = \frac{V^2}{Q_C} = \frac{120^2}{-1031.3} = -13.96\,\Omega$$

2.3.3 Summary of Basic Relationships in Inductive and Capacitive Circuits

In a sinusoidal steady state at a frequency f and at the corresponding angular frequency ω, the various relationships in circuits with L and C are summarized in Table 2.1.

TABLE 2.1 Summary of various relationships related to L and C in a sinusoidal steady-state circuit at an angular frequency ω

Relationship	L	C
Reactance	$X_L = \omega L$	$X_C = -\dfrac{1}{\omega C}$
Impedance	$Z_L = jX_L = j\omega L$	$Z_C = jX_C = j\left(-\dfrac{1}{\omega C}\right)$
Voltage and current	$\bar{V}_L = (j\omega L)\bar{I}_L$	$\bar{V}_C = j\left(-\dfrac{1}{\omega C}\right)\bar{I}_C$
Phase angle	Voltage leads current by 90 degrees $\phi = \phi_v - \phi_i = 90°$	Current leads voltage by 90 degrees $\phi = \phi_v - \phi_i = (-90°)$
Reactive power drawn	$Q_L = \dfrac{V_L^2}{X_L} = \text{positive}$ $Q_L = V_L I_L \sin(90°) = V_L I_L$	$Q_C = \dfrac{V_C^2}{X_C} = \text{negative}$ $Q_C = V_C I_C \sin(-90°) = -V_C I_C$

2.4 THREE-PHASE CIRCUITS

A basic understanding of three-phase circuits is essential to the study of electric power systems. Nearly all electricity, with only a few exceptions, is generated using three-phase AC generators. Figure 2.10 shows a simplified one-line diagram of a three-phase power system. Generated voltages (usually below 25 kV) are stepped-up using transformers to 230 kV to 500 kV levels for transferring power long distances over transmission lines from the generation site to load centers. At the receiving end of the transmission lines, near the load centers, these three-phase voltages are stepped down by transformers. Most motor loads above a few kW in power rating operate from three-phase voltages.

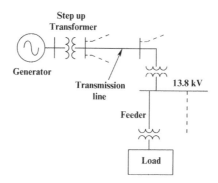

FIGURE 2.10 One-line diagram of a three-phase power system; the subtransmission is not shown.

Three-phase AC circuits are either Y or delta connected. We will investigate these under a sinusoidal steady state, balanced condition, which implies that all three voltages are equal in magnitude and displaced by 120 degrees ($2\pi/3$ radians) with respect to each other. The phase sequence of voltages is commonly assumed to be $a-b-c$, where the phase a voltage leads the phase b voltage by 120 degrees, and phase b leads phase c by 120 degrees ($2\pi/3$ radians), as shown in Figure 2.11. This applies to both the time-domain representation in Figure 2.11a and the phasor-domain representation in Figure 2.11b. Notice that in the $a-b-c$ sequence voltages plotted in Figure 2.11a, first v_{an} reaches its positive peak, then v_{bn} reaches its positive peak of $2\pi/3$ radians, and so on. We can represent these voltages in the phasor form in Figure 2.11b as

$$\overline{V}_{an} = V_{ph}\angle 0°, \quad \overline{V}_{bn} = V_{ph}\angle -120°, \quad \text{and} \quad \overline{V}_{cn} = V_{ph}\angle -240° \quad (2.24)$$

where V_{ph} is the rms phase-voltage magnitude and the phase a voltage is assumed to be the reference (with an angle of zero degrees).

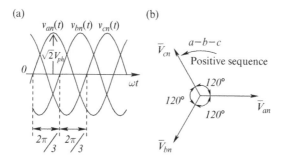

FIGURE 2.11 Three-phase voltages in the (a) time domain and (b) phasor domain.

For a balanced set of voltages given by Equation 2.24, at any instant, the sum of these phase voltages equals zero:

$$\bar{V}_{an} + \bar{V}_{bn} + \bar{V}_{cn} = 0 \rightarrow \text{and} \rightarrow v_{an}(t) + v_{bn}(t) + v_{cn}(t) = 0 \qquad (2.25)$$

2.4.1 Per-Phase Analysis in Balanced Three-Phase Circuits

A three-phase circuit can be analyzed on a per-phase basis, provided it has a balanced set of source voltages and equal impedances in each phase. Such a Y-connected circuit is shown in Figure 2.12a.

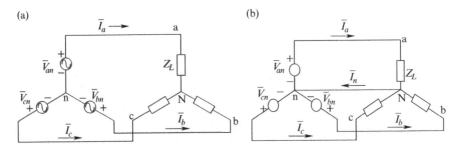

FIGURE 2.12 Balanced Y-connected, three-phase circuit. (a) Generic diagram; (b) source and load neutral connected by a zero-impedance wire.

In such a circuit, the source neutral n and the load neutral N are at the same potential. Therefore, hypothetically connecting these with a zero-impedance wire, as shown in Figure 2.12b, does not change the original three-phase circuit, which can now be analyzed on a per-phase basis. Selecting phase a for this analysis, the per-phase circuit is shown in Figure 2.13a.

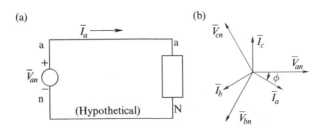

FIGURE 2.13 (a) Per-phase circuit; (b) the corresponding phasor diagram.

If $Z_L = |Z_L| \angle \phi$, using the fact that in a balanced three-phase circuit, phase quantities are displaced by 120 degrees with respect to each other, we find that

$$\bar{I}_a = \frac{\bar{V}_{an}}{Z_L} = \frac{V_{ph}}{|Z_L|}\angle{-\phi} \quad \bar{I}_b = \frac{\bar{V}_{bn}}{Z_L} = \frac{V_{ph}}{|Z_L|}\angle(-\phi - \frac{2\pi}{3}) \quad \bar{I}_c = \frac{\bar{V}_{cn}}{Z_L} = \frac{V_{ph}}{|Z_L|}\angle(-\phi - \frac{4\pi}{3})$$

$$(2.26)$$

The three-phase voltage and current phasors are shown in Figure 2.13b. The total real and reactive powers in a balanced three-phase circuit can be obtained by multiplying the per-phase values by a factor of three. The total power factor is the same as its per-phase value.

2.4.2 Per-Phase Analysis of Balanced Circuits Including Mutual Couplings

Most three-phase equipment such as generators, transmission lines, and motors consists of mutually coupled phases. For example, in a three-phase synchronous generator, the current in a phase winding produces flux lines that link not only that phase winding but also the other phase windings. In general, in a balanced three-phase circuit, this can be represented as shown in Figure 2.14a, where Z_{self} is the impedance of a phase by itself and Z_{mutual} represents mutual coupling. Therefore,

$$\bar{V}_{aA} = Z_{self}\bar{I}_a + Z_{mutual}\bar{I}_b + Z_{mutual}\bar{I}_c \tag{2.27}$$

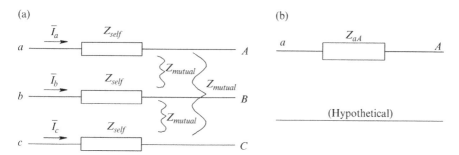

FIGURE 2.14 Balanced three-phase network with mutual couplings. (a) Three-phase representation; (b) per-phase representation.

In a balanced three-phase circuit under balanced excitation, three currents sum to zero: $\bar{I}_a + \bar{I}_b + \bar{I}_c = 0$. Therefore, using this condition in Equation 2.25,

$$\bar{V}_{aA} = (Z_{self} - Z_{mutual})\bar{I}_a = Z_{aA}\bar{I}_a \tag{2.28}$$

where

$$Z_{aA} = Z_{self} - Z_{mutual} \tag{2.29}$$

and the per-phase representation is as shown in Figure 2.14b.

2.4.3 Line-to-Line Voltages

In a balanced Y-connected circuit like that shown in Figure 2.12a, it is often necessary to consider the line-to-line voltages, such as those between phases *a* and *b*, and so on. Based on the previous analysis, we can refer to both neutral points *n* and *N* by a common term *n*, since the potential difference between *n* and *N* is zero. Therefore, in Figure 2.12a, as shown in the phasor diagram in Figure 2.15,

$$\overline{V}_{ab} = \overline{V}_{an} - \overline{V}_{bn}, \quad \overline{V}_{bc} = \overline{V}_{bn} - \overline{V}_{cn}, \quad \text{and} \quad \overline{V}_{ca} = \overline{V}_{cn} - \overline{V}_{an} \qquad (2.30)$$

Either using Equations 2.24 and 2.30 or working graphically from Figure 2.15, we can show the following, where V_{ph} is the rms magnitude of each of the phase voltages:

$$\overline{V}_{ab} = \sqrt{3}V_{ph}\angle\frac{\pi}{6}$$

$$\overline{V}_{bc} = \sqrt{3}V_{ph}\angle(\frac{\pi}{6} - \frac{2\pi}{3}) = \sqrt{3}V_{ph}\angle-\frac{\pi}{2} \qquad (2.31)$$

$$\overline{V}_{ca} = \sqrt{3}V_{ph}\angle(\frac{\pi}{6} - \frac{4\pi}{3}) = \sqrt{3}V_{ph}\angle-\frac{7\pi}{6}$$

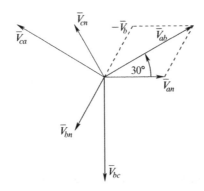

FIGURE 2.15 Line-to-line voltages in a three-phase circuit.

Comparing Equations 2.24 and 2.31, we see that the line-to-line voltages have an rms value that is $\sqrt{3}$ times the phase voltage (rms)

$$V_{LL} = \sqrt{3}V_{ph} \qquad (2.32)$$

and $\overline{V}_{ab}$ leads $\overline{V}_{an}$ by $\pi/6$ radians (30 degrees).

Example 2.6

In residential buildings where three-phase voltages are brought in, the rms value of the line-line voltage is $V_{LL} = 208\,\text{V}$. Calculate the rms value of the phase voltages.

Solution From Equation 2.32,

$$V_{ph} = \frac{V_{LL}}{\sqrt{3}} = 120\,\text{V}$$

2.4.4 Delta Connections in AC Machines and Transformers

So far, we have assumed that three-phase sources and loads are connected in a Y configuration, as shown in Figure 2.12a. However, in AC machines and transformers, the three-phase windings may be connected in a delta configuration. The relationship between the line currents and phase currents under a balanced condition is described in Appendix 2A.1.

2.4.4.1 Delta-Y Transformation of Load Impedances under Balanced Conditions

It is possible to replace the delta-connected load impedances with the equivalent Y-connected load impedances and vice versa. Under a totally balanced condition, the delta-connected load impedances in Figure 2.16a can be replaced by the equivalent Y-connected load impedances in Figure 2.16b, as shown in Figure 2.12a. We can then apply a per-phase analysis using Figure 2.13.

Under the balanced condition where all three impedances are the same, as shown in Figure 2.16, the impedance between terminals a and b, with the terminal c not connected, is $(2/3)Z_\Delta$ in the delta-connected circuit and $2Z_Y$ in the Y-connected circuit. Equating these two impedances,

$$Z_Y = \frac{Z_\Delta}{3} \tag{2.33}$$

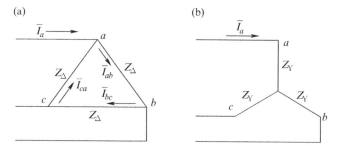

FIGURE 2.16 Delta-Y transformation of impedances under a balanced condition. (a) Delta representation; (b) Y representation.

The delta-Y transformation of unbalanced load impedances is analyzed in Appendix 2A.2.

2.4.5 Power, Reactive Power, and Power Factor in Three-Phase Circuits

We saw earlier that the per-phase analysis is valid for balanced three-phase circuits in a sinusoidal steady state. This implies that the real and reactive powers drawn by each phase are the same as if it were a single-phase load. Therefore, the total average real power and the reactive power in a three-phase circuit are ($V = V_{ph}$ and $I = I_{ph}$)

$$P_{\text{3-phase}} = 3 \times P_{\text{per-phase}} = 3VI \cos \phi \tag{2.34}$$

$$Q_{\text{3-phase}} = 3 \times Q_{\text{per-phase}} = 3VI \sin \phi \tag{2.35}$$

and the total apparent VA are

$$\left| S \right|_{\text{3-phase}} = 3 \times \left| S_{\text{per-phase}} \right| = 3VI \tag{2.36}$$

Therefore, the power factor in a three-phase circuit is the same as the per-phase power factor:

$$\text{power factor} = \frac{P_{\text{3-phase}}}{\left| S \right|_{\text{3-phase}}} = \frac{3VI \cos \phi}{3VI} = \cos \phi \tag{2.37}$$

However, there is one very important difference between three-phase and single-phase circuits in terms of instantaneous power. In both circuits, in each phase, the instantaneous power is pulsating, as given by Equation 2.11 and repeated here for phase a, where the phase current lags by a phase angle $\phi(t)$ behind the phase voltage, which is considered the reference phasor:

$$p_a(t) = \sqrt{2}V \cos \omega t \cdot \sqrt{2}I \cos(\omega t - \phi) = VI \cos \phi + VI \cos(2\omega t - \phi) \tag{2.38}$$

Similar expressions can be written for phases b and c:

$$\begin{aligned}
p_b(t) &= \sqrt{2}V \cos(\omega t - 2\pi/3) \cdot \sqrt{2}I \cos(\omega t - \phi - 2\pi/3) \\
&= VI \cos \phi + VI \cos(2\omega t - \phi - 4\pi/3)
\end{aligned} \tag{2.39}$$

$$\begin{aligned}
p_c(t) &= \sqrt{2}V \cos(\omega t - 4\pi/3) \cdot \sqrt{2}I \cos(\omega t - \phi - 4\pi/3) \\
&= VI \cos \phi + VI \cos(2\omega t - \phi - 8\pi/3)
\end{aligned} \tag{2.40}$$

Adding the three instantaneous powers in Equations 2.38 through 2.40 results in

$$p_{\text{3-phase}}(t) = 3VI \cos\phi = P_{\text{3-phase}} \qquad \text{(from Equation 2.34)} \qquad (2.41)$$

which shows that the combined three-phase power in the steady state is a constant equal to its average value, even on an instantaneous basis. This contrasts with the pulsating power in single-phase circuits, shown in Figure 2.7. The non-pulsating total instantaneous power in three-phase circuits results in non-pulsating torque in motors and generators and is the reason for preferring three-phase motors and generators over their single-phase counterparts.

Example 2.7
In the three-phase circuit shown in Figure 2.12a, $V_{LL} = 208\,\text{V}$, $|Z_L| = 10\,\Omega$, and the per-phase power factor is 0.8 (lagging). Calculate the capacitive reactance needed, in parallel with the load impedance in each phase, to make the power factor 0.95 (lagging).

Solution The three-phase circuit in Figure 2.12a can be represented by the per-phase circuit in Figure 2.13a. Assuming the input voltage as the reference phasor, $\bar{V}_{an} = \dfrac{208}{\sqrt{3}}\angle 0 = \underbrace{120}_{(=V)}\angle 0\,\text{V}$. The current $\bar{I}_L$ drawn by the load lags the voltage by an angle $\phi_L = \cos^{-1}(0.8) = 36.87°$, and

$$\bar{I}_L = \frac{V(=120)}{|Z_L|}\angle -\phi_L = 12\angle -36.87°\,\text{A}$$

Therefore, the per-phase real power P_L and the reactive var Q_L drawn by the load are

$$P_L = \frac{V^2}{|Z_L|}(\text{power factor}) = 1152\,\text{W}, \text{ and } Q_L = \frac{V^2}{|Z_L|}\sin\phi_L = 864\,\text{var}.$$

To make the net power factor 0.95 (lagging), a power-factor-correction capacitor of an appropriate value is connected in parallel with the load in each phase. Now the net current into the combination of the load impedance and the power-factor-correction capacitor is at an angle of $\phi_{\text{net}} = \cos^{-1}(0.95) = 18.195°$ (lagging). The net real power P_{net} drawn from the source still equals P_L: that is, $P_{\text{net}} = P_L$. Using the power triangle from Figure 2.8c, the net reactive var drawn from the source is

$$Q_{\text{net}} = P_{\text{net}} \tan(\phi_{\text{net}}) = 378.65\,\text{var}$$

Since

$$Q_{net} = Q_L - Q_{cap}$$

$$Q_{cap} = Q_L - Q_{net} = 864.0 - 378.65 = 485.35 \, \text{var}$$

Therefore, the capacitive reactance needed in parallel is

$$X_{cap} = \frac{V^2}{Q_{cap}} = 29.67 \, \Omega$$

2.5 REAL AND REACTIVE POWER TRANSFER BETWEEN AC SYSTEMS

In this course, it will be important to calculate the power flow between AC systems connected by transmission lines. Simplified AC systems can be represented by two AC voltage sources of the same frequency connected through a reactance X in series, as shown in Figure 2.17a, where the series resistance has been neglected for simplification.

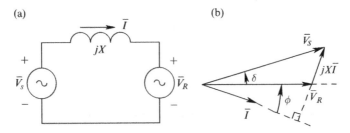

FIGURE 2.17 Power transfer between two AC systems. (a) Circuit diagram; (b) phasor diagram for the circuit in (a).

The phasor diagram for the system in Figure 2.17a, where the voltage $\bar{V}_R$ is assumed to be the reference voltage with a phase angle of zero, is shown in Figure 2.17b. Based on the load, the current may be at some arbitrary phase angle ϕ. In the circuit shown in Figure 2.17a,

$$\bar{I} = \frac{\bar{V}_S - \bar{V}_R}{jX} \tag{2.42}$$

At the receiving end, the complex power can be written as

$$S_R = P_R + jQ_R = V_R \bar{I}^* \tag{2.43}$$

Using the complex conjugate from Equation 2.42 into Equation 2.43,

$$P_R + jQ_R = V_R \left(\frac{V_S \angle(-\delta) - V_R}{-jX} \right) = \underbrace{\frac{V_S V_R \sin \delta}{X}}_{(=P_R)} + j \underbrace{\left(\frac{V_S V_R \cos \delta - V_R^2}{X} \right)}_{(=Q_R)} \quad (2.44)$$

$$\therefore \rightarrow P_R = \underbrace{\frac{V_S V_R}{X}}_{(=P_{max})} \sin \delta \rightarrow \text{where} \quad P_{max} = \frac{V_S V_R}{X} \rightarrow \quad (2.45)$$

which is the same as the sending end power P_S, assuming no transmission-line losses. This discussion shows that the real power P flows downhill on the phase angles of the voltages and not on their magnitudes. It is plotted in Figure 2.18 for positive values of δ.

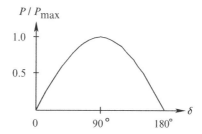

FIGURE 2.18   Power as a function of δ.

Focusing on the reactive power from Equation 2.44,

$$Q_R = \frac{V_S V_R \cos \delta}{X} - \frac{V_R^2}{X} \quad (2.46)$$

If the power transfer between the two systems is zero, then, from Equation 2.45, $\sin \delta$ and the angle δ are equal to zero. Under this condition, from Equation 2.46,

$$Q_R = \frac{V_S V_R}{X} - \frac{V_R^2}{X} = \frac{V_R}{X}(V_S - V_R) \rightarrow (\text{if } P_R = 0) \quad (2.47)$$

which shows that the reactive power at the receiving end is related to the difference $(V_S - V_R)$ between the two voltage magnitudes.

The real and reactive power transfers given by Equations 2.45–2.47 are extremely important in the discussion of power and reactive power flows in later chapters.

2.6   EQUIPMENT RATINGS, BASE VALUES, AND PER-UNIT QUANTITIES

2.6.1   Ratings

All power equipment, such as generators, transformers, and transmission lines, has a voltage rating V_{rated}, which is the nominal voltage at which it is designed to be operated based on its insulation or to avoid magnetic saturation at the operating frequency, whichever may be more limiting. Similarly, each piece of equipment has a current rating I_{rated}, which is the nominal current (usually specified in rms) at which it is designed to be operated in a steady state and beyond which excessive I^2R heating may cause damage to the equipment.

2.6.2   Base Values and Per-Unit Values

Equipment parameters are specified in per-unit quantities as fractions of the appropriate base values. Often, the rated voltage and rated current are chosen as base values, and the other base quantities are calculated based on them.

There are several reasons for specifying equipment parameters in per-unit quantities:

1. Regardless of the size of the equipment, per-unit values based on the equipment's voltage and current ratings lie in a narrow range and hence are easy to check or estimate.
2. Power systems often involve multiple transformers; such a system has several voltage and current ratings, as we will see in Chapter 6. In such a system, a set of base values is chosen that is common to the entire system. Using parameters in per-unit quantities of these common base values greatly simplifies the analysis: the voltage and current transformation due to the turns-ratio of the transformers are eliminated from calculations, as discussed in Chapter 5.

With V_{base} and I_{base}, the other base quantities can be calculated as follows:

$$R_{base}, X_{base}, Z_{base} = \frac{V_{base}}{I_{base}} \rightarrow (in\,\Omega) \tag{2.48}$$

$$G_{base}, B_{base}, Y_{base} = \frac{I_{base}}{V_{base}} \rightarrow (in\,\Omega) \tag{2.49}$$

$$P_{base}, Q_{base}, (VA)_{base} = V_{base}I_{base} \rightarrow (in\ watts,\ var,\ or\ VA) \tag{2.50}$$

In terms of these base quantities, the per-unit quantities can be specified as

$$\text{per-unit value} = \frac{\text{actual value}}{\text{base value}} \tag{2.51}$$

As mentioned earlier, equipment parameters are specified in per-unit quantities of the base values that usually equal its rated voltage and current. Another benefit of using per-unit quantities is that their values remain the same on per-phase basis or a total three-phase basis. For example, if we are considering per-phase power, the power in watts is one-third of the total power, and so is its base value compared to that in the three-phase case.

Example 2.8

In Example 2.7, calculate the per-unit values of the per-phase voltage, load impedance, load current, and load real and reactive powers if the line-line voltage base is 208 V (rms) and the base value of the three-phase power is 5.4 kW.

Solution Given that $V_{LL,base} = 208\,V$, $V_{ph,base} = 120\,V$. The base value of the per-phase power is $P_{ph,base} = \dfrac{5400}{3} = 1800\,W$. Therefore,

$$I_{base} = \frac{P_{ph,base}}{V_{ph,base}} = 15\,A \text{ and } Z_{base} = \frac{V_{ph,base}}{I_{base}} = 8\,\Omega$$

With these base values, $\bar{V}_{ph} = 1.0\angle 0\,pu$, $\bar{I} = 0.8\angle -36.87°\,pu$, $Z_L = 1.25\angle -36.87°\,pu$, $P_L = 0.64\,pu$, and $Q_L = 0.48\,pu$.

2.7 ENERGY EFFICIENCIES OF POWER SYSTEM EQUIPMENT

Power system equipment must be energy efficient and reliable, have a high power density (thus reducing its size and weight), and be low cost to make the overall system economically feasible. High energy efficiency is important for several reasons: it reduces operating costs by avoiding wasted energy, contributes less to global warming, and reduces the need for cooling, therefore increasing power density.

The energy efficiency η of the equipment shown in Figure 2.19 is

$$\eta = \frac{P_o}{P_{in}} \tag{2.52}$$

which, in terms of the output power P_o and the power loss P_{loss} within the equipment, is

$$\eta = \frac{P_o}{P_o + P_{loss}} \tag{2.53}$$

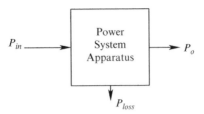

FIGURE 2.19   Energy efficiency $\eta = P_\mathrm{o} \, / \, P_\mathrm{in}$.

In power systems, pieces of equipment such as transformers and generators have percentage energy efficiencies in the upper 90s, and there are constant efforts to increase them further.

2.8   ELECTROMAGNETIC CONCEPTS

Much of the equipment used in power systems, including transformers, synchronous generators, transmission lines, and motor loads, requires a basic understanding of electromagnetic concepts, which will be reviewed in this section.

2.8.1   Ampere's Law

When a current i is passed through a conductor, a magnetic field is produced. The direction of the magnetic field depends on the direction of the current. As shown in Figure 2.20a, the current through a conductor, perpendicular and *into* the plane of the paper, is represented by $\times$; this current produces a magnetic field in a clockwise direction. Conversely, the current *out of* the plane of the paper, represented by a dot, produces a magnetic field in a counterclockwise direction, as shown in Figure 2.20b.

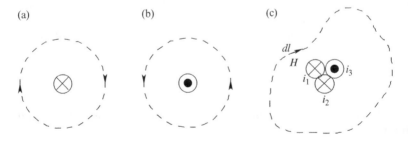

FIGURE 2.20   (a) Clockwise magnetic field by current going into the paper; (b) counterclockwise magnetic field by current coming out of the paper; (c) Ampere's Law.

The magnetic-field intensity H produced by current-carrying conductors can be obtained using Ampere's law, which in its simplest form states that, at any time, the line (contour) integral of the magnetic field intensity along *any* closed path equals the total current enclosed by this path. Therefore, in Figure 2.20c,

$$\oint H d\ell = \sum i \qquad (2.54)$$

where $\oint$ represents a contour or closed-line integration. Note that the scalar H in Equation 2.54 is the component of the magnetic field intensity (a vector field) in the direction of the differential length $d\ell$ along the closed path. Alternatively, we can express the field intensity and differential length as vector quantities, which requires a dot product on the left side of Equation 2.54.

Example 2.9
Consider the coil in Figure 2.21, which has $N = 25$ turns; the toroid has an inside diameter $ID = 5 \, \text{cm}$ and an outside diameter $OD = 5.5 \, \text{cm}$.

(a)   (b)

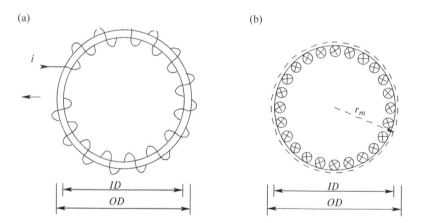

FIGURE 2.21   Coil for Example 2.9. (a) Toroid with a coil; (b) ampere-turns enclosed by the magnetic field.

For a current $i = 3 \, \text{A}$, calculate the field intensity H along the mean-path length within the toroid.

Solution Due to symmetry, the magnetic field intensity H_m along a circular contour within the toroid is constant. In Figure 2.21, the mean radius $r_m = \dfrac{1}{2}\left(\dfrac{OD + ID}{2}\right)$. Therefore, the mean path of length $\ell_m (= 2\pi r_m = 0.165 \, \text{m})$ encloses the current i N times, as shown in Figure 2.21b. Therefore, from Ampere's law in Equation 2.54, the field intensity along this mean path is

$$H_m = \frac{Ni}{\ell_m} \qquad (2.55)$$

which for the given values can be calculated as

$$H_m = \frac{25 \times 3}{0.165} = 454.5 \, \text{A/m}$$

If the width of the toroid is much smaller than the mean radius r_m, it is reasonable to assume a uniform H_m throughout the cross-section of the toroid.

The field intensity in Equation 2.55 has units of A/m; note that *turns* is a unit-less quantity. The product Ni is commonly referred to as the *ampere-turns* or *mmf* F that produces the magnetic field. The current in Equation 2.55 may be DC or time-varying. If the current is time-varying, the relationship in Equation 2.55 is valid on an instantaneous basis; that is, $H_m(t)$ is related to $i(t)$ by N / ℓ_m.

2.8.2   Flux Density B and the Flux ϕ

At any instant of time t for a given H-field, the density of flux lines, called the flux density B (in units of T for teslas), depends on the permeability μ of the material on which this H-field is acting. In air,

$$B = \mu_o H \rightarrow \mu_o = 4\pi \times 10^{-7} \left[\frac{\text{henries}}{\text{m}} \right] \qquad (2.56)$$

where μ_o is the permeability of air or free space.

2.8.3   Ferromagnetic Materials

Ferromagnetic materials guide magnetic fields and, due to their high permeability, require small ampere-turns (a small current for a given number of turns) to produce the desired flux density. These materials exhibit the multivalued nonlinear behavior shown by their *B-H* characteristics in Figure 2.22a. Imagine that the toroid in Figure 2.21 consists of a ferromagnetic material such as silicon steel. If the current through the coil is slowly varied in a sinusoidal manner with time, the corresponding H-field will cause one of the hysteresis loops shown in Figure 2.22a to be traced. Completing the loop once results in a net dissipation of energy within the material. This energy loss per cycle is referred to as *hysteresis loss*.

Increasing the peak value of the sinusoidally varying H-field results in a bigger hysteresis loop. Joining the peaks of the hysteresis loop, we can approximate the *B-H* characteristic by the single curve shown in Figure 2.22b. At low values of the magnetic field, the *B-H* characteristic is assumed to be linear with a constant slope, such that

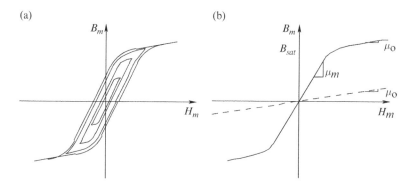

FIGURE 2.22 *B-H* characteristics of ferromagnetic materials. (a) Actual; (b) approximation.

$$B_m = \mu_m H_m \tag{2.57a}$$

where μ_m is the permeability of the ferromagnetic material. Typically, the μ_m of a material is expressed in terms of a permeability μ_r relative to the permeability of air:

$$\mu_m = \mu_r \mu_o \rightarrow \left(\mu_r = \frac{\mu_m}{\mu_o} \right) \tag{2.57b}$$

In ferromagnetic materials, μ_m can be several thousand times larger than μ_o.

In Figure 2.22b, the linear relationship (with a constant μ_m) is approximately valid until the knee of the curve is reached, beyond which the material begins to saturate. Ferromagnetic materials are often operated up to a maximum flux density slightly above the knee of 1.6–1.8 T, beyond which many more ampere-turns are required to increase flux density only slightly. In the saturated region, the incremental permeability of the magnetic material approaches μ_o, as shown by the slope of the curve in Figure 2.22b.

In this course, we will assume that the magnetic material is operating in its linear region, and therefore its characteristic can be represented by $B_m = \mu_m H_m$, where μ_m remains constant.

2.8.4 Flux ϕ

Magnetic flux lines form closed paths, as shown in Figure 2.23's toroidal magnetic core, which is surrounded by the current-carrying coil.

The flux in the toroid can be calculated by selecting a circular area A_m in a plane perpendicular to the direction of the flux lines. As discussed in Example 2.9, it is reasonable to assume a uniform H_m and hence a uniform

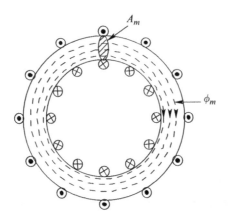

FIGURE 2.23   Toroid with flux ϕ_m.

flux-density B_m throughout the core cross-section. Substituting for H_m from Equations 2.55 into 2.57a,

$$B_m = \mu_m \frac{Ni}{\ell_m} \tag{2.58}$$

where B_m is the density of flux lines in the core. Therefore, assuming a uniform B_m, the flux ϕ_m can be calculated as

$$\phi_m = B_m A_m \tag{2.59}$$

where flux has units of webers (Wb). Substituting for B_m from Equations 2.58 into 2.57,

$$\phi_m = A_m \left(\mu_m \frac{Ni}{\ell_m} \right) = \frac{Ni}{\underbrace{\left(\dfrac{\ell_m}{\mu_m A_m} \right)}_{\mathfrak{R}_m}} \tag{2.60}$$

where Ni equals the ampere-turns (or mmf F) applied to the core, and the term in brackets on the right side is called the *reluctance* $\mathfrak{R}_m$ of the magnetic core. From Equation 2.60,

$$\mathfrak{R}_m = \frac{\ell_m}{\mu_m A_m} [\text{A/Wb}] \tag{2.61}$$

Equation 2.60 makes it clear that the reluctance has units A/Wb. Equation 2.61 shows that the reluctance of a magnetic structure – for example, the toroid in

Figure 2.23 – is linearly proportional to its magnetic path length and inversely proportional to both its cross-sectional area and the permeability of its material.

Equation 2.60 shows that the amount of flux produced by the applied ampere-turns $F (= Ni)$ is inversely proportional to the reluctance $\Re$; this relationship is analogous to Ohm's law $(I = V / R)$ in electric circuits in a DC steady state.

2.8.5 Flux Linkage

If all the turns of a coil like the one in Figure 2.23 are linked by the same flux ϕ, then the coil has a flux linkage λ, where

$$\lambda = N\phi \tag{2.62}$$

In the absence of magnetic saturation, the flux linkage λ is related to the coil current i by the coil inductance, as illustrated in the next section.

2.8.6 Inductances

At any instant of time in the coil shown in Figure 2.24a, the flux linkage of the coil (due to flux lines entirely in the core) is related to the current i by a parameter defined as the inductance L_m

$$\lambda_m = L_m i \tag{2.63}$$

where the inductance $L_m(= \lambda_m / i)$ is constant if the core material is in its linear operating region. The coil inductance in the linear magnetic region can be calculated by multiplying all the factors shown in Figure 2.24b, which are based on earlier equations:

$$L_m = \left(\frac{N}{\ell_m}\right)\mu_m A_m N = \frac{N^2}{\left(\dfrac{\ell_m}{\mu_m A_m}\right)} = \frac{N^2}{\Re_m} \tag{2.64}$$

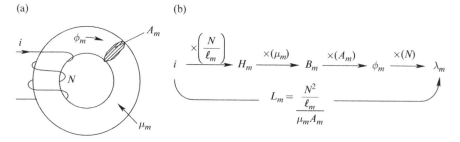

FIGURE 2.24 (a) Coil inductance; (b) visual mathematical interpretation of (a).

Equation 2.64 indicates that the inductance L_m is strictly a property of the magnetic circuit (i.e. the core material, the geometry, and the number of turns), provided that the operation is in the linear range of the magnetic material, where the slope of its *B-H* characteristic can be represented by a constant μ_m.

Example 2.10
In the rectangular toroid shown in Figure 2.25, $w = 5\,\text{mm}$, $h = 15\,\text{mm}$, the mean path length $\ell_m = 18\,\text{cm}$, $\mu_r = 5000$, and $N = 100$ turns. Calculate the coil inductance L_m, assuming that the core is unsaturated.

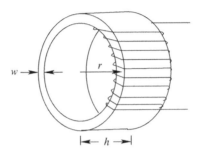

FIGURE 2.25   Rectangular toroid for Example 2.10.

Solution From Equation 2.61,

$$\mathfrak{R}_m = \frac{\ell_m}{\mu_m A_m} = \frac{0.18}{5000 \times 4\pi \times 10^{-7} \times 5 \times 10^{-3} \times 15 \times 10^{-3}} = 38.2 \times 10^4 \frac{\text{A}}{\text{Wb}}$$

Therefore, from Equation 2.64,

$$L_m = \frac{N^2}{\mathfrak{R}_m} = 26.18 \text{ mH}$$

2.8.7   Faraday's Law: Induced Voltage in a Coil Due to the Time Rate of Change of the Flux Linkage

In our discussion so far, we have established in magnetic circuits relationships between the electrical quantity i and the magnetic quantities H, B, ϕ, and λ. These relationships are valid under DC (static) conditions and at any instant when these quantities vary with time. We will now examine the voltage across the coil under time-*varying* conditions. In the coil shown in Figure 2.26, Faraday's law dictates that the time rate of change of the flux linkage equals the voltage across the coil at any instant:

$$e(t) = \frac{d}{dt}\lambda(t) = N\frac{d}{dt}\phi(t) \tag{2.65}$$

This assumes that all flux lines link all N-turns such that $\lambda = N\phi$. The polarity of the emf $e(t)$ and the direction of $\phi(t)$ in Equation 2.65 are yet to be justified.

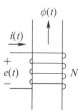

FIGURE 2.26 Voltage polarity and direction of flux and current.

This relationship is valid no matter what is causing the flux to change. One possibility is that a second coil is placed on the same core. When the second coil is supplied by a time-varying current, mutual coupling causes the flux ϕ through the coil shown in Figure 2.26 to change with time. The other possibility is that a voltage $e(t)$ is applied across the coil in Figure 2.26, causing a change in flux that can be calculated by integrating both sides of Equation 2.65 with respect to time

$$\phi(t) = \phi(0) + \frac{1}{N}\int_0^t e(\tau)\cdot d\tau \tag{2.66}$$

where $\phi(0)$ is the initial flux at $t = 0$ and τ is a variable of integration.

Recalling the Ohm's law equation $v = Ri$, the current direction through a resistor is defined to be into the terminal chosen to be of the positive polarity. This is the passive sign convention. Similarly, in the coil in Figure 2.26, we can establish the voltage polarity and the flux direction in order to apply Faraday's law, given by Equations 2.65 and 2.66. If the flux direction is given, we can establish the voltage polarity as follows: first determine the direction of a hypothetical current that will produce flux in the same direction as given. Then the positive polarity for the voltage is at the terminal this hypothetical current is entering. Conversely, if the voltage polarity is given, imagine a hypothetical current entering the positive-polarity terminal. Based on how the coil is wound – for example, in Figure 2.26 – this current determines the flux direction for use in Equations 2.65 and 2.66. Following these rules to determine the voltage polarity and the flux direction is easier than applying Lenz's law.

Another way to determine the polarity of the induced emf is to apply Lenz's law, which states the following: if a current is allowed to flow due to the voltage induced by an increasing flux linkage, for example, the direction of this hypothetical current will be to oppose the flux change.

Example 2.11
In the structure shown in Figure 2.26, the flux $\phi_m (= \hat{\phi}_m \sin \omega t)$ linking the coil is varying sinusoidally with time, where $N = 300$ turns, $f = 60\,\text{Hz}$, and the cross-sectional area $A_m = 10\,\text{cm}^2$. The peak flux density $\hat{B}_m = 1.5\,\text{T}$. Calculate the expression for the induced voltage with the polarity shown in Figure 2.26. Plot the flux and induced voltage as functions of time.

Solution From Equation 2.59, $\hat{\phi}_m = \hat{B}_m A_m = 1.5 \times 10 \times 10^{-4} = 1.5 \times 10^{-3}$ Wb.

From Faraday's law in Equation 2.63, the induced voltage below is

$$e(t) = \omega N \hat{\phi}_m \cos \omega t = 2\pi \times 60 \times 300 \times 1.5 \times 10^{-3} \times \cos \omega t = 169.65 \cos \omega t \text{ V}$$

The induced voltage and the flux waveform are plotted in Figure 2.27, from which we can conclude that the induced voltage phasor leads the flux phasor by 90 degrees.

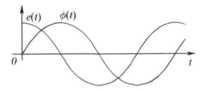

FIGURE 2.27   Plot for Example 2.11.

Example 2.11 illustrates that the voltage is induced due to $d\phi / dt$, regardless of whether any current flows in that coil.

2.8.8   Leakage and Magnetizing Inductances

Just as conductors guide currents in electric circuits, magnetic cores guide *flux* in *magnetic circuits*. But there is an important difference. In electric circuits, the conductivity of copper is approximately 10^{20} times higher than that of air, allowing leakage currents to be neglected at DC or low frequencies such as 60 Hz. In magnetic circuits, however, the permeabilities of magnetic materials are only around 10^4 times greater than that of air. Because of this relatively low ratio, not all of the flux is confined to the core in the structure shown in Figure

2.28a, and the core window also has flux lines called *leakage*. Note that the coil in Figure 2.28a is drawn schematically; in practice, the coil consists of multiple layers, and the core is designed to fit as snugly to the coil as possible, thus minimizing the unused window area.

(a)   (b)

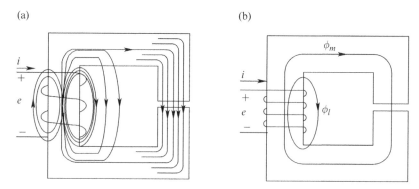

FIGURE 2.28 (a) Flux flow in a coil including leakage flux; (b) approximate model for (a).

The leakage effect makes accurate analysis of magnetic circuits more difficult, so it requires sophisticated numerical methods, such as finite element analysis. However, we can account for the effect of leakage fluxes by making certain approximations. We can divide the total flux ϕ into two parts: the magnetic flux ϕ_m, which is completely confined to the core and links all N turns, and the leakage flux, which is partially or entirely in air and is represented by an equivalent leakage flux ϕ_ℓ, which also links all N turns of the coil but does not follow the entire magnetic path, as shown in Figure 2.28b. Thus,

$$\phi = \phi_m + \phi_\ell \qquad (2.67)$$

where ϕ is the equivalent flux that links all N turns. Therefore, the total flux linkage of the coil is

$$\lambda = N\phi = \underbrace{N\phi_m}_{\lambda_m} + \underbrace{N\phi_\ell}_{\lambda_\ell} = \lambda_m + \lambda_\ell \qquad (2.68)$$

The total inductance (called the *self-inductance*) can be obtained by dividing both sides of Equation 2.68 by the current i:

$$\underbrace{\frac{\lambda}{i}}_{L_{\text{self}}} = \underbrace{\frac{\lambda_m}{i}}_{L_m} + \underbrace{\frac{\lambda_\ell}{i}}_{L_\ell} \qquad (2.69)$$

Therefore,

$$L_{\text{self}} = L_m + L_\ell \tag{2.70}$$

where L_m is often called the *magnetizing inductance* due to ϕ_m in the magnetic core, and L_ℓ is called the *leakage inductance* due to the leakage flux ϕ_ℓ. From Equation 2.70, the total flux linkage of the coil can be written as

$$\lambda = (L_m + L_\ell)i \tag{2.71}$$

Hence, from Faraday's law in Equation 2.65,

$$e(t) = L_\ell \frac{di}{dt} + \underbrace{L_m \frac{di}{dt}}_{e_m(t)} \tag{2.72}$$

This results in the circuit shown in Figure 2.29a. In Figure 2.29b, the voltage drop due to the leakage inductance can be shown separately so that the voltage induced in the coil is solely due to the magnetizing flux. The coil resistance R can then be added in series to complete the representation of the coil.

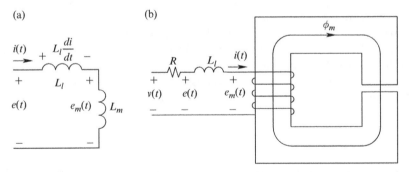

FIGURE 2.29 Analysis including the leakage flux. (a) Circuit for Figure 2.28b; (b) leakage flux represented as leakage inductance.

REFERENCE

1. Any textbook covering basic electric circuits and electromagnetic field theory.

PROBLEMS

2.1   Express the following voltages as phasors: (a) $v_1(t) = \sqrt{2} \times 100 \cos(\omega t - 30°)$ V and (b) $v_2(t) = \sqrt{2} \times 100 \cos(\omega t + 30°)$ V.

2.2   The series *R-L-C* circuit shown in Figure 2.3a is in a sinusoidal steady state at a frequency of 60 Hz. $V = 120$ V, $R = 1.5\ \Omega$, $L = 20$ mH, and $C = 100\ \mu F$. Calculate $i(t)$ in this circuit by solving the differential Equation 2.3.

2.3   Repeat Problem 2.2 using the phasor-domain analysis.

2.4   In a linear circuit in a sinusoidal steady state with only one active source $\overline{V} = 90\angle 30°$ V, the current in a branch is $\overline{I} = 5\angle 15°$A. Calculate the current in the same branch if the source voltage were $100\angle 0°$V.

2.5   If a voltage $100\angle 0°$V is applied to the circuit shown in Example 2.1, calculate P, Q, and the power factor. Show that $Q = \sum_{k} I_k^2 X_k$.

2.6   Show that a series-connected load with R_s and X_s, as shown in Figure P2.6a, can be represented by a parallel combination, as shown in Figure 2.6b, where $R_p = (R_s^2 + X_s^2)/ R_s$ and $X_p = (R_s^2 + X_s^2)/ X_s$.

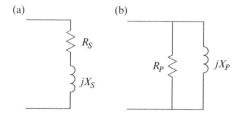

FIGURE P2.6   (a) Series-connected *R* and *X*; (b) equivalent representation of (a) using a parallel connection.

2.7   Using the results of the series-parallel conversion in Problem 2.6, calculate the equivalent R_P and X_P to represent the circuit from Example 2.2.

2.8   Confirm the calculations of the equivalent R_P and X_P in Problem 2.7 by using the *P* and *Q* calculated in Example 2.3, recognizing that in the parallel representation, *P* is entirely associated with R_P and *Q* entirely with X_P.

2.9   In Example 2.5, calculate the compensating capacitor in parallel necessary to make the overall power factor 0.9 (leading).

2.10   An inductive load connected to a 120 V (rms), 60 Hz AC source draws 5 kW at a power factor of 0.8. Calculate the capacitance required in parallel with the load to bring the combined power factor to 0.95 (lagging).

2.11   A positive sequence (*a-b-c*), balanced, Y-connected voltage source has a phase *a* voltage given as $\overline{V}_a = \sqrt{2} \times 100\angle 30°$ V. Obtain the time-domain voltages $v_a(t)$, $v_b(t)$, $v_c(t)$, and $v_{ab}(t)$, and show all of these as phasors.

2.12   A balanced three-phase inductive load is supplied in a steady state by a balanced, Y-connected, three-phase voltage source with a phase voltage of 120 V rms. The load draws a total of 10 kW at a power factor of 0.9. Calculate the rms value of the phase currents and the magnitude of the per-phase load impedance, assuming a Y-connected load. Draw a phasor diagram showing all three voltages and currents.

2.13   Repeat Problem 2.12, assuming a balanced delta-connected load.

2.14   The balanced circuit in Figure 2.14 shows the impedance of three-phase cables connecting the source terminals (a, b, c) to the load terminals (A, B, C), where $Z_{self} = (0.3 + j1.5)\Omega$ and $Z_{mutual} = j0.5\Omega$. Calculate $\bar{V}_A$ if $\bar{V}_a = 1000\angle 0°\,V$ and $\bar{I}_a = 10\angle -30°\,A$, where $\bar{V}_a$ and $\bar{V}_A$ are voltages with respect to a common neutral.

2.15   Similar to the calculation in Section 2.4.2, in the balanced circuit shown in Figure P2.15, calculate the per-phase capacitive currents in terms of the voltages, capacitances, and frequency.

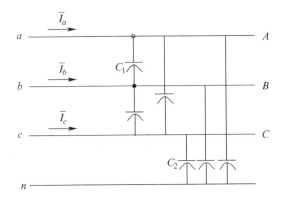

FIGURE P2.15   Balanced circuit with capacitance for Problem 2.15.

2.16   In the per-phase circuit shown in Figure 2.17a, the power transfer per-phase is 1 kW from side 1 to side 2. $V_S = 100\,V$, $\bar{V}_R = 95\angle 0°\,V$, and $X = 1.5\Omega$. Calculate the current, the phase angle of $\bar{V}_S$, and the per-phase Q_R supplied to the receiving end.

2.17   In a radial system represented by the circuit shown in Figure 2.17a, $X = 1.5\Omega$. Consider the source voltage constant at $\bar{V}_S = 100\angle 0°$. Calculate and plot V_S / V_R if the load varies in a range from 0 to 1 kW at the following three power factors: unity, 0.9 (lagging), 0.9 (leading).

2.18   Repeat Example 2.8 if the base three-phase power is changed to 3.6 kW but all else remains the same.

2.19   Repeat Example 2.8 if the baseline line voltage is changed to 240 V but all else remains the same.

2.20   Repeat Example 2.8 if the base line-line voltage is changed to 240 V *and* the base three-phase power is changed to 3.6 kW.

2.21   In Example 2.9, calculate the field intensity within the core a) very close to the inside diameter and b) very close to the outside diameter. c) Compare the results with the field intensity along the mean path.

2.22   In Example 2.9, calculate the reluctance in the path of flux lines if $\mu_r = 2000$.

2.23   Consider a core with the dimensions given in Example 2.9. The coil requires an inductance of 25 μH. The maximum current is 3 A, and the maximum flux density is not to exceed 1.3 T. Calculate the number of turns N and the relative permeability μ_r of the magnetic material that should be used.

2.24   In Problem 2.23, assume the permeability of the magnetic material to be infinite. To satisfy the conditions of maximum flux density and the desired inductance, a small air gap is introduced. Calculate the length of this air gap (neglecting the effect of flux fringing) and the number of turns N.

APPENDIX 2A

2A.1   Line and Phase Currents in a Delta-Connected Load under Balanced Conditions

Figure 2A.1 shows a balanced delta-connected load being supplied by a balanced three-phase source. From Kirchhoff's current law,

$$\bar{I}_a = \bar{I}_{ab} - \bar{I}_{ca} \tag{2A.1}$$

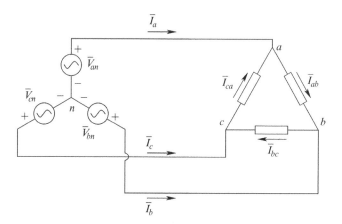

FIGURE 2A.1   Balanced delta-connected load.

Let $\bar{I}_{ab} = I_{\text{phase}} \angle 0°$. Since the system is balanced, $\bar{I}_{ca} = I_{\text{phase}} \angle -240°$. Therefore, from Equation (2A.1), as shown in Figure 2A.2,

$$\bar{I}_a = \bar{I}_{ab} - \bar{I}_{ca} = \sqrt{3}\,I_{\text{phase}}\angle -30° \tag{2A.2}$$

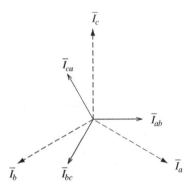

FIGURE 2A.2 Current phasors in a balanced delta-connected load.

Figure 2A.2 shows that the line-current magnitudes are $\sqrt{3}$ times those of the currents within the delta-connected load.

2A.2 Transformation Between Delta- And Wye-Connected Impedances

Consider the impedances connected in Figure 2A.3, where in general they may be unbalanced.

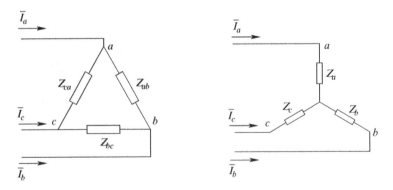

FIGURE 2A.3 Delta and Y configurations.

To arrive at the appropriate transformation, consider that node c is disconnected from the rest of the circuit, i.e. $\bar{I}_c = 0$. Since both configurations are equivalent as far as the external circuit is concerned, the impedance between nodes a and b must be the same. Therefore,

$$Z_a + Z_b = Z_{ab} \parallel (Z_{ca} + Z_{bc}) \tag{2A.3}$$

Similarly, considering nodes a and b to be open, respectively,

$$Z_c + Z_a = Z_{ca} \parallel (Z_{bc} + Z_{ab}) \tag{2A.4}$$

$$Z_b + Z_c = Z_{bc} \parallel (Z_{ab} + Z_{ca}) \tag{2A.5}$$

Solving these equations simultaneously,

$$Z_a = \frac{Z_{ab} Z_{ca}}{Z_{ab} + Z_{bc} + Z_{ca}} \tag{2A.6}$$

$$Z_b = \frac{Z_{bc} Z_{ab}}{Z_{ab} + Z_{bc} + Z_{ca}} \tag{2A.7}$$

$$Z_c = \frac{Z_{ca} Z_{bc}}{Z_{ab} + Z_{bc} + Z_{ca}} \tag{2A.8}$$

For the reverse transform, from Equations (2A.6)–(2A.8),

$$Z_a Z_b = \frac{Z_{ab}^{\,2} Z_{bc} Z_{ca}}{\left(Z_{ab} + Z_{bc} + Z_{ca}\right)^2} \tag{2A.9}$$

$$Z_b Z_c = \frac{Z_{ab} Z_{bc}^{\,2} Z_{ca}}{\left(Z_{ab} + Z_{bc} + Z_{ca}\right)^2} \tag{2A.10}$$

$$Z_c Z_a = \frac{Z_{ab} Z_{bc} Z_{ca}^{\,2}}{\left(Z_{ab} + Z_{bc} + Z_{ca}\right)^2} \tag{2A.11}$$

Adding these three equations,

$$Z_a Z_b + Z_b Z_c + Z_c Z_a = \frac{Z_{ab}^{\,2} Z_{bc} Z_{ca} + Z_{ab} Z_{bc}^{\,2} Z_{ca} + Z_{ab} Z_{bc} Z_{ca}^{\,2}}{\left(Z_{ab} + Z_{bc} + Z_{ca}\right)^2} \tag{2A.12}$$

and

$$\frac{1}{Z_a}(Z_aZ_b + Z_bZ_c + Z_cZ_a) = \frac{1}{Z_a}\frac{Z_{ab}{}^2 Z_{bc}Z_{ca} + Z_{ab}Z_{bc}{}^2 Z_{ca} + Z_{ab}Z_{bc}Z_{ca}{}^2}{(Z_{ab} + Z_{bc} + Z_{ca})^2} \quad (2A.13)$$

Substituting for Z_a from Equation (2A.6) on the right side of the previous equation,

$$\frac{(Z_aZ_b + Z_bZ_c + Z_cZ_a)}{Z_a} = \frac{(Z_{ab} + Z_{bc} + Z_{ca})}{Z_{ab}Z_{ca}} \times$$

$$\frac{Z_{ab}{}^2 Z_{bc}Z_{ca} + Z_{ab}Z_{bc}{}^2 Z_{ca} + Z_{ab}Z_{bc}Z_{ca}{}^2}{(Z_{ab} + Z_{bc} + Z_{ca})^2} \quad (2A.14)$$

The right side in Equation (2A.14) simplifies to Z_{bc}, and therefore

$$Z_{bc} = \frac{(Z_aZ_b + Z_bZ_c + Z_cZ_a)}{Z_a} \quad (2A.15)$$

By symmetry,

$$Z_{ab} = \frac{(Z_aZ_b + Z_bZ_c + Z_cZ_a)}{Z_c} \quad (2A.16)$$

and

$$Z_{ca} = \frac{(Z_aZ_b + Z_bZ_c + Z_cZ_a)}{Z_b} \quad (2A.17)$$

3

ELECTRIC ENERGY AND THE ENVIRONMENT

3.1 INTRODUCTION

It is difficult to get information about the generation and consumption of electricity worldwide. Therefore, we have concentered on the electricity scene in the United States, using the information in [1]. In 2021, the net generation of electricity from utility-scale generators in the United States was about 4,116 billion kilowatt-hours (kWh) (or about 4.12 trillion kWh). Figure 3.1 shows the generating capacity in 2021 from major sources.

Figure 3.2 shows U.S. electricity generation by major energy sources from 1950 to 2021.

The percentage shares of utility-scale electricity generation by major energy sources in 2021 were natural gas, 38.3%; coal, 21.8%; nuclear, 18.9%; renewables (total), 20.1%; nonhydroelectric renewables, 13.8%; and hydroelectric, 6.3%.

3.2 CHOICES AND CONSEQUENCES

As illustrated in Figure 3.2, electric energy is derived from the following sources. Most are from the sun directly or indirectly (except nuclear energy), which dates back approximately 6.6 billion years:

- Hydro
- Fossil fuels: coal, natural gas, oil

Electric Power Systems with Renewables: Simulations Using PSS®E, Second Edition. Ned Mohan and Swaroop Guggilam.
© 2023 John Wiley & Sons, Inc. Published 2023 by John Wiley & Sons, Inc.
Companion Website: www.wiley.com/go/mohaneps

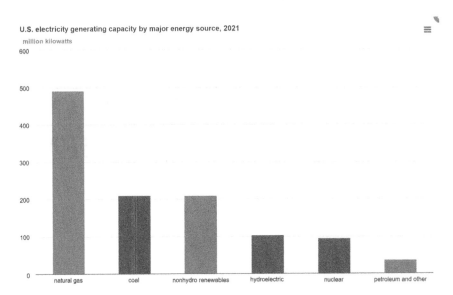

FIGURE 3.1   U.S. electricity generation capacity from major sources in 2021.

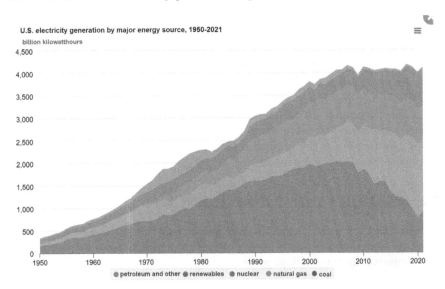

FIGURE 3.2   U.S. electricity generation by major energy sources, 1950–2021.

- Nuclear
- Renewables: wind, solar, biomass and geothermal

There are environmental consequences from using energy; in fact, all human activities have an environmental impact. These consequences are as follows:

- Greenhouse gases, primarily carbon dioxide
- Sulfur dioxide
- Nitrogen oxides
- Mercury
- Thermal pollution

3.3   HYDROPOWER

This is one of the oldest ways to produce electricity. In hydropower plants, as shown in Figure 3.3, water in a reservoir or behind a dam in a river is discharged through penstocks into turbines that spin generators and produce electricity. In such power plants, turbine efficiencies can be better than 93% [2].

Hydropower can be classified based on the available *head* (drop in the river level): high, medium, or low (run of river). The higher the head (*H* in Figure 3.3), the greater the potential energy that can be converted by the turbine into mechanical output. Because the environmental impact and social implications can be enormous, including uprooting people, high- and medium-head hydro are generally not considered renewable energy, although technically they are. On the other hand, low-head (run-of-river) hydro, where the kinetic energy associated with flowing water is converted by a turbine into mechanical output, is considered renewable. As an example, in a recent study, heads below 66 feet (20 m) were considered renewable energy. Hydropower plants can respond very quickly – in a matter of a few seconds – by changing their output to meet changes in the power demand.

 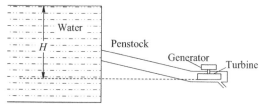

FIGURE 3.3   Hydropower. Source: www.bpa.gov.

3.4   FOSSIL-FUEL-BASED POWER PLANTS

Fossil fuels are the main source of electric energy worldwide except in a few countries such as Norway (hydro) and France (nuclear). These fossil fuels are primarily in the following form:

- Coal
- Natural gas
- Oil

Fossil fuels are derived from the sun's energy by decaying vegetation and plants. In this sense, fossil fuels are also renewable but at a time scale that is much longer – hundreds of millions of years – than human existence. All these sources of energy generation are discussed briefly in the following sections.

3.4.1   Coal-Fired Power Plants

As mentioned earlier, coal-fired power plants are the main source of electricity in most countries. The United States is richly endowed with this resource, with 30% of the world's total – enough to last hundreds of years. However, burning coal leads to serious environmental consequences: greenhouse gases, as discussed later in this chapter. The available coal can be divided into the following categories, where each type has its own characteristics in terms of energy content and resulting pollution: anthracite, bituminous, sub-bituminous, lignite, and peat.

Coal is burned using various coal-firing mechanisms with different efficiencies and carbon emissions into the atmosphere: mechanical stokers, pulverized-coal firing, cyclone-furnace firing, fluidized-bed combustion, and gasification. The heat from burning coal produces steam from water, which is used in the Rankine thermodynamic cycle shown in Figure 3.4, where water is used as a working fluid. Heat is added to the water at high pressure in a boiler to produce steam, which expands through the turbine blades and is then cooled in a condenser. The thermal efficiency of such a cycle is typically in the 35–40% range: for example, 9,000 BTU/kWh is typical, corresponding to a conversion efficiency of approximately 38%.

In the discussion of efficiencies, we should note that the limit of any thermal cycle efficiency is less than that of the Carnot cycle [3] efficiency η_c

$$\eta_C = \frac{T_H - T_L}{T_H} \tag{3.1}$$

where the temperatures are in Kelvin, T_H is the higher temperature at which heat is added, and T_L is the lower temperature at which heat is rejected.

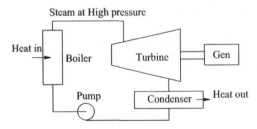

FIGURE 3.4   Rankine thermodynamic cycle in coal-fired power plants.

Coal-fired power plants have enormous environmental consequences, as discussed later in this chapter, but the availability of coal at an affordable cost makes it an attractive choice. Coal plants take a long time to bring online from a cold start, and changing their power output on demand from one level to another takes minutes.

3.4.2 Natural-Gas and Oil Power Plants

Natural gas is plentiful in certain regions of the world. Gas plants do not have the same environmental consequences as coal-fired power plants, such as mercury pollution, but natural gas is a hydrocarbon-based fuel that also contributes to greenhouse gases. Gas plants are relatively inexpensive and quick to build and can have reasonable efficiencies, as discussed shortly. Oil power plants are similar to gas-fired power plants in their operation and efficiencies.

3.4.2.1 Single-Cycle Gas Turbines

Single-cycle gas plants use the Brayton thermodynamic cycle. The Brayton thermodynamic cycle can take many forms, the simplest of which is shown in Figure 3.5: natural gas is burned in a combustion chamber in the presence of compressed air and then expanded through turbine blades like steam in the Rankine cycle.

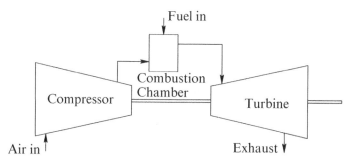

FIGURE 3.5 Brayton thermodynamic cycle in natural-gas power plants.

In single-cycle gas-fired power plants, efficiencies are around 35%, which is satisfactory when the primary function is to supply peak loads.

3.4.2.2 Combined-Cycle Gas Turbines

In combines-cycle gas turbines, heat in the exhaust of the Brayton cycle is recovered to operate another steam-based Rankine cycle, and the efficiency can be boosted to a range of 55 to 60%. Because of their high efficiencies, combined-cycle power plants can be base loaded, and the capacity of such plants can be as great as 500 MW.

3.5 NUCLEAR POWER

As mentioned earlier, the nuclear energy in matter dates back to the time of creation, the so-called Big Bang, such that a gram of completely fissionable material can produce 1000 MW-day of energy! A large nuclear plant can save 50,000 barrels of oil per day, which at the present cost of more than $60 per barrel amounts to over $3 million per day. Nuclear power doesn't contribute to greenhouse gases, but it creates a serious problem of storing radioactive waste for generations to come. This problem has not been resolved in the United States, where there is a de facto moratorium on constructing new nuclear power plants.

The nuclear process results in the release of energy in a nuclear reactor, which produces steam to spin a turbine and a generator connected to it, similar to that in coal-fired power plants using the Rankine thermodynamic cycle. Nuclear processes can be broadly classified into two categories that are briefly examined in the following sections: fusion and fission.

3.5.1 Nuclear Fusion

Fusion is a thermonuclear reaction similar to the reaction that occurs incessantly on the sun, where hydrogen atoms fuse together and release energy in the process. These atoms are at extremely high temperatures for this reaction to take place. In fusion reactors, two deuterium (H^2 or D) atoms fuse, resulting in a tritium (H^3 or T) atom and a proton (p), and in the process release 4 MeV of energy, where 1 MeV equals 4.44×10^{-20} kWh:

$$D + D \rightarrow T + p + 4\,\mathrm{Mev} \tag{3.2}$$

Unlike fission reactors (discussed in the next section), fusion reactors have a smaller radioactive waste problem. However, the plasma formed from fusing atoms at millions of degrees must be confined away from the reactor walls. This problem is unlikely to be solved in the near future at a commercial scale, although another attempt is under way [4].

3.5.2 Nuclear Fission Reactors

All large commercial nuclear reactors rely on the fission process. In this process, neutrons strike atoms of fissionable material like uranium, "splitting" them and, in the process, releasing more neutrons to keep this chain reaction continuing and releasing energy. This reaction can be expressed as follows

$$_{92}U^{235} + {}_0n^1 \rightarrow {}_{54}Xe^{140} + {}_{38}Sr^{94} + 2{}_0n^1 + 196\,\mathrm{Mev} \tag{3.3}$$

where subscripts correspond to the atomic number and superscripts to the atomic mass. This reaction shows that a neutron striking a $_{92}U^{235}$ atom results in two other fission products, xenon $_{54}Xe^{140}$ and strontium $_{38}Sr^{94}$, and two neutrons and releases 196 MeV of energy. The resulting neutrons strike other uranium atoms and keep the chain reaction going in a controlled fashion.

In the fission process, the neutrons produced are at very high energy and, unless slowed, would fail to strike other uranium atoms to keep the chain reaction going. Therefore, they must be "moderated," often by coolant that transports heat and acts as a moderator. This chain reaction is controlled by control rods that consist of, for example, boron that absorbs a neutron and thus allows control over the rate at which the chain reaction is maintained. Generally, nuclear plants are base loaded, implying that they are operated at their rated capacity. In case of emergency, a nuclear reactor can be shut down by releasing a chemical shim that acts as a "poison" that "kills" the nuclear reaction. However, using a chemical shim may result in the reactor being inoperable for the days required to clean away its effect.

3.5.2.1 Pressurized Water Reactors (PWRs)
In the past, a few boiling-water reactors (BWRs), as shown in Figure 3.6a, were built, in addition to a few gas-cooled reactors (GCRs). But most modern reactors are pressurized-water reactors (PWRs). In PWRs, as shown in Figure 3.6b,

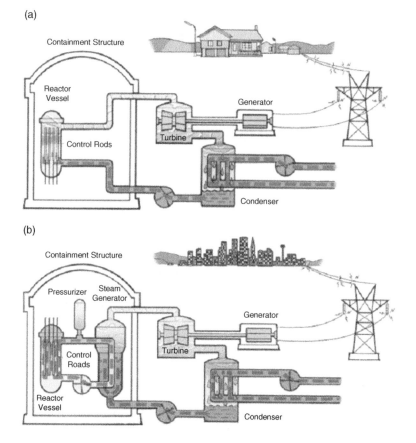

FIGURE 3.6 (a) Boiling-water reactor and (b) pressurized-water reactor [5].

water that is under pressure to keep it from boiling acts as a coolant to transport heat and as a moderator to continue the chain reaction. Using a heat exchanger, the secondary cycle results in the water boiling. The steam produced is used in a Rankine thermodynamic cycle, similar to that in coal-fired plants, to spin the turbine and the generator.

3.5.2.2 Pressurized Heavy Water Reactors (PHWRs)

Natural uranium is composed of less than 1% U^{235} and approximately 99% U^{238}. In PWRs and BWRs, where natural water is used as the moderator, enriched uranium must be used so that it has a greater percentage of U^{235} compared to natural uranium. However, enriching uranium requires centrifuges that are extremely energy expensive and difficult to justify without a large weapons program. For example, in countries like Canada, CANDU reactors utilize natural uranium but use heavy water as the moderator and coolant. Heavy water is derived from natural water in such a way it has a heavier isotope of hydrogen present. Except for this difference, PHWRs are similar in principle to PWRs.

3.5.3.3 Fast Breeder Reactors

These reactors utilize plutonium and produce more nuclear fuel than they consume. Therefore, they can be self-sustaining; however, their extensive use of plutonium, a substance for making nuclear weapons, makes them an extremely risky proposition, particularly in this post-9/11 world. With its large Superphénix reactors, France has made the largest implementation of such breeder reactors.

3.6 RENEWABLE ENERGY

Renewable energy is a means of reducing the amount of pollution caused by fossil-fuel-based power plants. Renewable sources include wind, solar, biomass, and others. Some of these are described briefly.

3.6.1 Wind Energy

Wind energy is an indirect manifestation of solar energy caused by uneven heating of the earth's surface by the sun. Of all the renewable energies, wind has come a long way, and this potential is just beginning to be realized. Figure 3.7 shows the wind potential in states in the United States, where several areas have good to excellent wind conditions.

In wind, the mass flow rate $\dot{m}$ in kg/s is

$$\dot{m} = \rho A V \qquad (3.4)$$

where ρ is the wind density in kg/m^3, A is the cross-sectional area in m^2 perpendicular to the wind velocity, and V is the wind velocity in m/s.

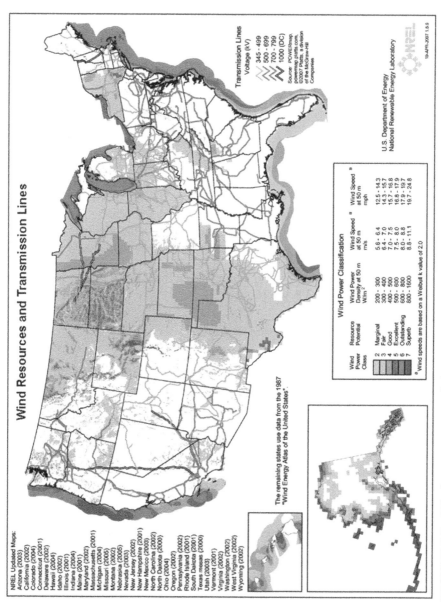

FIGURE 3.7 Wind-resource map of the United States [6].

Wind power is the rate of the kinetic energy of the wind stream

$$P_{tot} = \frac{1}{2}\dot{m}V^2 = \frac{1}{2}\rho A V^3 \qquad (3.5)$$

which shows the highly nonlinear dependence of wind power on the cube of the wind velocity. All this power cannot be removed from the wind, or it will "pile up" behind the turbine, which is impossible. The power that can be derived is the total power times a coefficient of performance, C_p, which is the ratio of the power available in the wind to that harnessed:

$$P_w = C_p P_{tot} = C_p \left(\frac{1}{2}\rho A V^3 \right) \qquad (3.6)$$

A detailed derivation in [3] shows that in the limit, theoretically, $C_{p,max} = 0.5926$. As wind approaches the plane in which the turbine blades are rotating, the pressure builds up, and the velocity begins to go down. After the plane of the blades, which is not very thick, the pressure begins to decrease and equals the pressure that exists much before the windmill. However, the wind-stream velocity keeps falling and levels off at a value lower than the initial velocity due to the energy extracted from it.

This characteristic C_p in Equation 3.6 is a function of the tip-speed ratio λ, as plotted in Figure 3.8, where

$$\text{blade tip-speed ratio } \lambda = \frac{\omega_m r}{V} \qquad (3.7)$$

in which r is the radius of the turbine blades in m, ω_m is the turbine rotational speed in rad/s, and V is the wind speed in m/s.

In practice, the maximum attainable value of C_p is generally around 0.45 at a pitch-angle of $\theta = 0$ degrees (this angle is zero when the blades are vertical to the wind direction). As shown in Figure 3.8, for each blade pitch-angle θ, C_p reaches a maximum at a particular value of the tip-speed ratio λ, which, for a given wind speed V, can be obtained by controlling the turbine rotational speed ω_m. The curves for various values of the pitch-angle θ shows that the power harnessed from wind can be regulated by controlling the pitch-angle of the blades and thus "spilling" some of the wind at very high wind speeds to prevent the power output from exceeding its design (rated) value.

3.6.1.1 Types of Generation Schemes in Windmills

Commonly used schemes for power generation in windmills require a gearing mechanism because the wind turbine rotates at very slow speeds, whereas the generator operates at a high speed close to the synchronous speed, which at the 60 Hz line frequency would be 1800 rpm for a four-pole and 900 rpm for an eight-pole machine. Therefore, the nacelle contains a gearing mechanism that boosts the turbine speed to drive the generator at a higher speed; the need for a

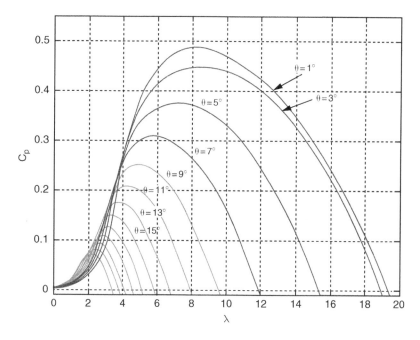

FIGURE 3.8 C_p as a function of λ [7]; these vary based on the turbine design.

gearing mechanism is one of the inherent drawbacks of such schemes. In very large windmills used for offshore applications, there are efforts to use direct-drive (without gears) wind turbines; however, most windmills use gearing. This subsection describes various types of wind-generation schemes.

Induction Generators, Directly Connected to the Grid. As shown in Figure 3.9, this is the simplest scheme, where a wind-turbine-driven squirrel-cage induction generator is directly connected to the grid through a back-to-back connected thyristor pair for a soft start. Therefore, it is the least expensive and uses a rugged squirrel-cage rotor induction machine.

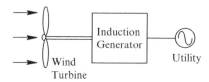

FIGURE 3.9   Induction generator directly connected to the grid [8].

For the induction machine to operate in its generator mode, the rotor speed must be greater than the synchronous speed. The drawback of this scheme is that since the induction machine always operates very close to the synchronous speed, it is not possible to achieve the optimum coefficient of performance, C_p, at all wind speeds, and hence is not optimum at low and high wind speeds compared to variable speed schemes described later.

Another disadvantage of this scheme is that a squirrel-cage induction machine always operates at a lagging power factor (that is, it draws reactive power from the grid as an inductive load would). Therefore, a separate source, such as shunt-connected capacitors, is often needed to supply the reactive power to overcome the lagging power factor operation of the induction machine.

Doubly Fed, Wound-Rotor Induction Generators. The scheme in Figure 3.10 utilizes a wound-rotor induction machine where the stator is directly connected to the utility supply. The rotor is injected with desired currents through a power-electronics interface, which is discussed in Chapter 7. Typically, four-fifths of the power flows directly from the stator to the grid, and only one-fifth of the power flows through the power electronics in the rotor circuit. The drawback of this scheme is that it uses a wound-rotor induction machine where the currents to the three-phase wound rotor are supplied through slip-rings and brushes, which require maintenance. Even though power electronics are expensive, since they are rated only one-fifth of the system rating, the overall cost is not much higher than the previous scheme. However, there are several distinct advantages compared to the previous scheme.

The scheme using a doubly fed wound-rotor induction machine can typically operate in a range ±30% around the synchronous speed, and hence it is able to capture more power at lower and higher wind speeds compared to the previous scheme because it can operate at above, as well as below, the synchronous speed. It can also supply reactive power, whereas, in the previous scheme, the squirrel-cage induction machine only absorbs reactive power. Therefore, the scheme using a doubly fed wound-rotor induction machine is quite popular in windmills being installed in the United States.

Power Electronics Connected Generator. In the third scheme, shown in Figure 3.11, a squirrel-cage induction generator or a permanent-magnet generator is

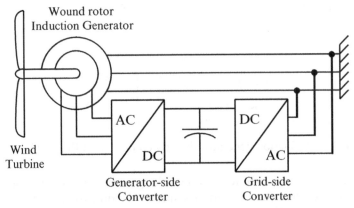

FIGURE 3.10   Doubly fed, wound-rotor induction generator [8, 9].

connected to the grid through a power electronics interface (described in chapter 7). This interface consists of two converters. The converter at the generator end supplies the reactive power excitation needed if it is an induction generator. Its frequency of operation is controlled to be optimum for the prevailing wind speed. The converter at the line end is capable of absorbing or supplying reactive power in a continuous manner. This is the most flexible arrangement, using a rugged squirrel-cage machine or a high-efficiency permanent-magnet generator that can operate in a very wide wind-speed range, and is the likely contender for future arrangements as the cost of the power electronics interface that handles the entire power output of the system is continuing to fall.

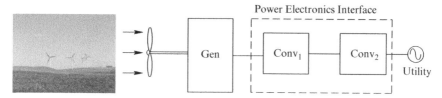

FIGURE 3.11   Power electronics connected generator [10].

3.6.1.2   Challenges in Harnessing Wind Energy

Wind power has enormous potential. North Dakota and South Dakota alone can potentially supply two-thirds of the present electric energy needs of the United States. But wind power also has many challenges. Wind is variable, and its power varies as the cube of the wind velocity. Therefore, it is difficult to use as a conventional dispatchable source by energy control centers. To overcome the dispatchability problem, research is being conducted on storage: for example, in flywheels for a short duration, supplemented by other means of generation, such as biodiesel when the wind dies down for longer periods. The other problem with wind resources is that they are located far from load centers, and harnessing this energy requires new transmission lines to be built.

3.6.2   Photovoltaic Energy

As mentioned earlier, except for nuclear, all forms of energy resources are indirectly based on solar energy. Photovoltaic (PV) cells convert the energy in the sun's rays directly into electricity. Solar energy is the ultimate distributed resource; however, the cost of photovoltaic cells is a barrier preventing its economic feasibility, and a breakthrough is awaited.

PV cells consist of a *pn* junction where incident photons in the sun's rays cause excess electrons and holes to be generated above their normal thermal equilibrium. This causes a potential to be developed, and if the external circuit is completed, it results in the flow of electrons and thus the flow of current. The $v-i$ characteristic of a photovoltaic cell is shown in Figure 3.12, which shows that each cell produces an open-circuit voltage of approximately 0.6 V. The

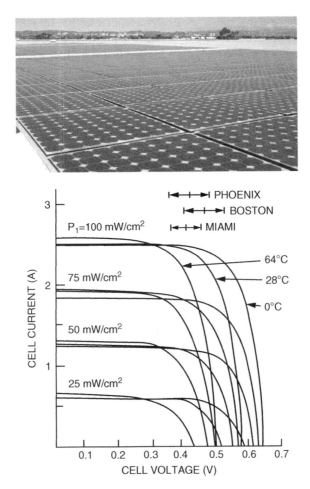

FIGURE 3.12 Photovoltaic cell characteristics [11].

short-circuit current is limited, and the power available is at its optimum value around the "knee" of the $v-i$ characteristic.

In PV systems, the PV arrays (typically four of them connected in series) provide a voltage of 53–90 V DC, which the power electronic system, as shown in Figure 3.13, converts to 120 V/60 Hz sinusoidal voltage suitable for inter-facing with the single-phase utility. A maximum-power-point tracking circuit in Figure 3.13 allows these cells to be operated at this maximum-power point.

There are several types of PV cells, like mono-crystalline silicon with an efficiency of 15–18%, multi-crystalline silicon with an efficiency of 13–18%, and amorphous silicon with an efficiency of 5–8%. As mentioned earlier, the cost of any of these panels at present is economically prohibitive, and a breakthrough is needed to make them feasible for power production, such as using windmills.

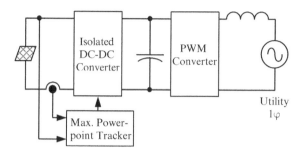

FIGURE 3.13   Photovoltaic systems.

3.6.3   Fuel Cells

Lately, there has been a great deal of interest in and effort devoted to fuel cell systems. Fuel cells use hydrogen (and possibly other fuels) as the input and, through a chemical reaction, directly produce electricity with water and heat as byproducts. In fuel cells, hydrogen reacts with a catalyst to produce electrons and protons. The protons pass through a membrane. The electrons flow through the external circuit and combine with the protons and oxygen to produce water. Heat is generated in this process as a byproduct. Therefore, fuel cells have none of the environmental consequences of conventional fuels, and their efficiency can be as high as 60%.

Fuel-cell output is a DC voltage, as shown in Figure 3.14, and therefore the need for power electronics converters to interface with the utility is the same in fuel-cell systems as in PV systems.

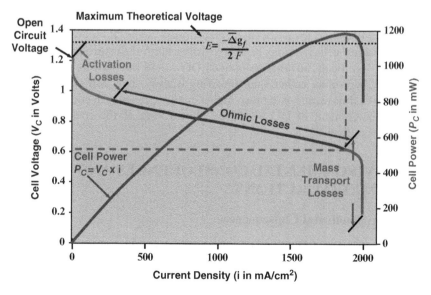

FIGURE 3.14   Fuel cell $v - i$ relationship and cell power [12].

At present, fuel cells are not fully competitive, and several challenges must be overcome. It is not clear what the best way would be to produce and transport hydrogen. If some other fuel, such as natural gas, was used, what would be the economic feasibility of reformers that would be needed in conjunction with fuel cells? However, intense research is being conducted on the development of various fuel cells. Some of them are listed here, with the expected percentage system efficiency in brackets: polymer electrolyte membrane (PEM) (32–40%), alkaline (32–45%), phosphoric acid (36–45%), molten carbonate (43–55%), and solid oxide (43–55%).

3.6.4 Biomass

Various other sources are under consideration for electrical generation. Biomass consists of biological materials like agricultural waste, wood, and manure. Other possibilities are vegetable oil, ethanol from corn and other crops, and gas from anaerobic decomposition. Biomass includes refuse-derived fuel. The use of biomass and its derivatives for commercial power production is just beginning, but significant effort is being put into this field.

3.7 DISTRIBUTED GENERATION (DG)

As mentioned in Chapter 1, the utility landscape is changing, and although in its infancy, there is growing use of distributed generation (DG). This type of generation, as the name implies, is distributed and usually much smaller in power rating than conventional power plants. Therefore, DG is often spurred by renewable resources such as windmills. In addition, there is a movement to generate electricity local to the load, to minimize the cost of transmission and distribution lines and the associated power losses. Such DG may be by microturbines and fuel cells that may be able to utilize natural gas through a reformer. One of the significant advantages of such DG would be to utilize the heat produced as a byproduct rather than throwing it away, as is common in central power plants, thus resulting in much higher energy efficiency in comparison. An ultimate achievement in DG would be photovoltaic if the cost of PV cells decreased significantly.

3.8 ENVIRONMENTAL CONSEQUENCES AND REMEDIAL ACTIONS

3.8.1 Environmental Consequences

Burning fossil fuels in power plants produces greenhouse gases, such as CO_2, that lead to climate change. The greenhouse effect of CO_2 is illustrated in Figure 3.15. Similar to the glass panes of a greenhouse, radiation from the sun in the form of ultraviolet light passes through the atmosphere, with some of it

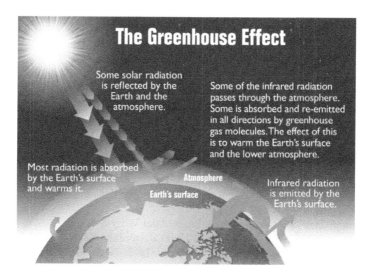

FIGURE 3.15 Greenhouse effect [13].

absorbed and some reflected. The radiation that passes through is mostly absorbed by the earth, warming it. Heat is radiated back, but due to the low temperature of the earth's surface, this radiation is infrared. Again, like the glass in a greenhouse, greenhouse gases such as carbon dioxide and nitrous oxide in the atmosphere trap the infrared heat, so less is radiated through and back to space. This greenhouse effect is leading to climate change, with disastrous consequences predicted.

There are other consequences as well. Burning fossil fuels results in increased concentrations of sulfur and nitrogen oxides in the atmosphere. These are converted into acid rain or acid snow, with risks to health and aquatic life. Certain types of coal lead to mercury pollution that is extremely harmful to health and becomes concentrated in fish, for example. Since the efficiency of the thermodynamic cycle in fossil fuel plants is only around 40%, the rest of the energy goes directly into creating thermal pollution of the water used to cool it. This can have adverse consequences, such as the growth of algae, etc.

3.8.2 Remedial Actions

Remedial actions include flue-gas desulfurization, scrubbers for nitrogen oxide removal, and electrostatic participators. Such remedies increase the cost of power plants and hence the generated electricity. Carbon sequestering is also being investigated.

Photovoltaics on the residential, commercial, and utility scales have enormous potential, and further research into making them more economical is being aggressively pursued.

REFERENCES

1. U.S. Department of Energy. www.eia.doe.gov.
2. H. M. Rustebakke (ed). 1983. *Electric Utility Systems and Practices*, 4th ed. John Wiley & Sons.
3. M. M. El-Wakil. 1984. *Powerplant Technology*. McGraw-Hill Companies.
4. *Technology Review Magazine*. September 2005. MIT.
5. United States Nuclear Regulatory Commission. http://www.nrc.gov/reading-rm/basic-ref/students/reactors.html.
6. National Renewable Energy Lab. www.nrel.gov.
7. K. Clark, N. W. Miller, and J. J. Sanchez-Gasca. 2009. "Modeling of GE Wind Turbine-Generators for Grid Studies." GE Energy Report, Version 4.4.
8. N. Mohan (ed). 2011. *Electric Machines and Drives: A First Course*. John Wiley & Sons.
9. N. Mohan. 2001. *Advanced Electric Drives*. John Wiley & Sons.
10. N. Mohan. 2011. *Power Electronics: A First Course*. John Wiley & Sons.
11. N. Mohan, T. Undeland, and W. P. Robbins. 1995. *Power Electronics: Converters, Applications and Design*. John Wiley & Sons.
12. National Energy Technology Laboratory. www.NETL.DOE.gov.
13. U.S. Environmental Protection Agency. http://www.epa.gov/climatechange/kids/index.html.

PROBLEMS

3.1 What are the sources of electric energy?

3.2 What are the environmental consequences of using electric energy?

3.3 Describe hydropower plants.

3.4 Describe two nuclear processes. Which one is used for commercial nuclear power production?

3.5 Describe different types of nuclear reactors.

3.6 Describe coal-fired power plants.

3.7 Describe gas-fired power plants.

3.8 What is the potential and what are the challenges associated with wind energy?

3.9 How does wind-turbine efficiency depend on the tip-speed ratio and the pitch-angle of the blades?

3.10 What are the three most commonly used generation schemes associated with harnessing wind energy?

3.11 What is the principle underlying the operation of photovoltaic cells?

3.12 What is the principle underlying the operation of fuel cells?

3.13 What is the greenhouse effect?

3.14 What are other environmental consequences of power production from using fossil fuels?

3.15 What are the present and potential remedies to minimize the environmental impacts of using fossil fuels?

3.16 A U.S. Department of Energy (DOE) report estimates that over 122 billion kWh/year can be saved in the manufacturing sector in the United

States by using mature and cost-effective conservation technologies. Calculate (a) how many 1000 MW generating plants are needed to operate constantly to supply this wasted energy and (b) the annual savings in dollars if the cost of electricity is $0.10 /kWh.

3.17   Electricity generation in the United States in the year 2000 was approximately 3.8×10^9 MW-hrs, 16% of which was used for heating, ventilating, and air conditioning. Based on a DOE report, as much as 30% of the energy can be saved in such systems by using adjustable-speed drives. On this basis, calculate the savings in energy per year and relate that to 1000 MW generating plants needed to operate constantly to supply this wasted energy.

3.18   The total amount of electricity that could potentially be generated from wind in the United States has been estimated at 10.8×10^9 MW-hrs annually. If one-tenth of this potential was developed, estimate the number of 2 MW windmills that would be required, assuming that, on average, a windmill produces only 25% of the energy it is capable of.

3.19   In problem 3.18, if each 2 MW windmill has 20% of its output power flowing through the power electronics interface, estimate the total rating of these power electronics interfaces in kW.

3.20   Lighting in the United States consumes 19% of the generated electricity. Compact fluorescent lamps (CFLs) consume one-fourth of the power consumed by incandescent lamps for the same light output. Electricity generation in the United States in the year 2000 was approximately 3.8×10^9 MW-hrs. Based on this information, estimate the savings in MW-hrs annually, assuming all lighting at present is by incandescent lamps, which will be replaced by CFLs.

3.21   Fuel-cell systems that also utilize the heat produced can achieve efficiencies approaching 80%, more than double that of gas-turbine-based electrical generation. Assume that 25 million households produce an average of 5 kW. Calculate the percentage of electricity generated by these fuel-cell systems compared to the annual electricity generation in the United States of 3.8×10^9 MW-hrs.

3.22   Induction cooking based on power electronics is estimated to be 80% efficient compared to 55% for conventional electric cooking. If an average home consumes 2 kW-hrs daily using conventional electric cooking, and 50 million households switch to induction cooking in the United States, calculate the annual savings in electricity usage.

3.23   Assume the average energy density of sunlight to be 800 W/m^2 and the overall photovoltaic system efficiency to be 10%. Calculate the land area covered with photovoltaic cells needed to produce 1,000 MW, the size of a typical large central power plant.

3.24   In problem 3.23, the solar cells are distributed on top of roofs, each with an area of 50 m^2. Calculate the number of homes needed to produce the same power.

4

AC TRANSMISSION LINES AND UNDERGROUND CABLES

4.1 NEED FOR TRANSMISSION LINES AND CABLES

Electricity is often generated in areas that are remote from load centers like metropolitan areas. Transmission lines form an important link in the power system structure to transport large amounts of electrical power with as little power loss as possible, keeping the system operating stably and at a minimum cost. Transmission-line access has become an important bottleneck in power systems operation today, with the building of additional transmission lines becoming a serious hurdle to overcome. This is one of the challenges facing the large-scale harnessing of wind energy, for example.

Most transmission systems consist of AC overhead transmission lines. Although we will mostly discuss overhead AC transmission lines, the analysis presented in this chapter applies to underground AC cables as well, as described briefly, which are used near metropolitan areas.

Several high-voltage DC (HVDC) transmission systems are also used to transport large amounts of power over long distances. Such systems are expected to be used more frequently. HVDC systems require power electronic converters and thus are discussed in Chapter 7, along with other applications of power electronics in power systems.

Electric Power Systems with Renewables: Simulations Using PSS®E, Second Edition. Ned Mohan and Swaroop Guggilam.
© 2023 John Wiley & Sons, Inc. Published 2023 by John Wiley & Sons, Inc.
Companion Website: www.wiley.com/go/mohaneps

4.2   OVERHEAD AC TRANSMISSION LINES

The discussion presented here applies to transmission as well as distribution lines, although the focus is on transmission lines. Transmission lines always consist of three phases at commonly used voltages of 115 kV, 161 kV, 230 kV, 345 kV, 500 kV, and 765 kV. Voltages lower than 115 kV are generally considered subtransmission and distribution voltages. Voltages higher than 765 kV are being considered for long-distance transmission.

A typical 500 kV transmission tower is shown in Figure 4.1a. There are three phases (*a*, *b*, and *c*), where each phase consists of three conductor bundles hung from the steel tower through insulators, as shown in Figure 4.1b. Conductors with multiple stands of aluminum on the outside and steel at the core, as shown in Figure 4.1c, are called aluminum conductor steel reinforced (ACSR). The current through the conductor primarily flows through the aluminum, which has conductivity more than seven times that of steel and is used at the core to provide tensile strength and thus prevent the lines from excessive sagging between towers. Typically, there are approximately five towers per mile.

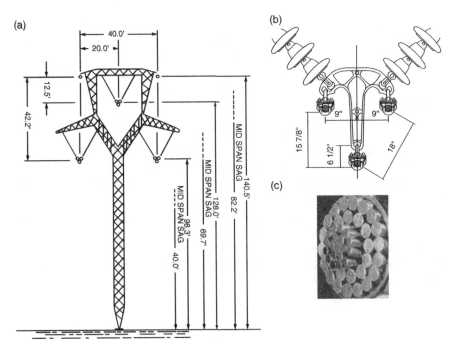

FIGURE 4.1   (a) Example of a 500 kV transmission tower; (b) each phase with three conductor bundles; (c) inside the conductor. **Source:** University of Minnesota EMTP course.

The parameters of a transmission line depend on the voltage level for which it is designed since higher voltage lines require larger values for the separation between conductors, height of the conductors, and conductor separation from the tower that is grounded. Proper clearance is needed to keep electric field strengths at the ground level under the transmission lines within the limit, which is specified to be 8 kV (rms)/m by the Minnesota Environmental Quality Board, for example.

4.2.1   Shield Wires, Bundling, and Cost

4.2.1.1   Shield Wires
As shown in Figure 4.1a, shield wires are often used to shield transmission-line phase conductors from being struck by lightning. These are also called ground wires, periodically grounded through the tower. Therefore, it is important to achieve as small a tower footing resistance as possible.

4.2.1.2   Bundling
As shown in Figure 4.1b, bundled conductors are used to minimize electric field strength at the conductor surface to avoid the corona effect discussed later in this chapter. According to [1], the maximum surface conductor gradient should be less than 16 kV/cm. Bundling conductors also reduces the line series inductance and increases the shunt capacitance, both of which are beneficial in loading lines to higher power levels. However, conductor bundling results in higher costs and increases the clearance required from the tower [2]. At line voltages lower than 345 kV, most transmission lines consist of single conductors. At 345 kV, most transmission lines consist of bundled conductors: usually a two-conductor bundle at a spacing of 18 in. At 500 kV, transmission lines use bundled conductors, such as the three-conductor bundle shown in Figure 4.1b, with a spacing of 18 in.

4.2.1.3   Cost
The cost of such lines depends on a variety of factors, but as a rough estimate, a 345 kV line costs from $500,000 per mile (1.6 km) in rural areas to over $2 million per mile in urban areas of Minnesota, with an average of approximately $750,000 per mile. Similar rough estimates are available for other transmission voltages.

4.3   TRANSPOSITION OF TRANSMISSION-LINE PHASES

In the structure shown in Figure 4.1a, the phases are arranged in a triangular fashion. In other structures, these phases may be all horizontal or all vertical, as shown in Figure 4.2a. It is clear from these arrangements that the electrical and magnetic couplings are not equal between the phases. For example, in Figure 4.2a, the coupling between phases *a* and *b* differs from that between phases *a* and *c*. To make these three phases appear balanced, they are transposed, as shown in

Figure 4.2b. This unbalance is not of much significance for short lines: for example, less than 100 km in length. But in long lines, the length of the barrel, equal to three sections or one cycle of transposition, as shown in Figure 4.2b, is recommended to be approximately 150 km for a triangular configuration and less for horizontal or vertical arrangements [3]. Although transmission lines are seldom transposed, we will assume the three phases to be perfectly balanced in our simplified analysis to follow.

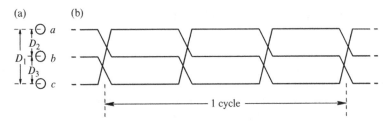

FIGURE 4.2 Transposition of transmission lines. (a) Conductor positions; (b) conductor transpositions.

4.4 TRANSMISSION-LINE PARAMETERS

Transmission-line resistances, inductances, and capacitances are distributed throughout the length of the line. Assuming the three phases to be balanced, we can easily calculate the transmission-line parameters on a per-phase basis, as shown in Figure 4.3, where the bottom conductor is neutral (in general, hypothetical) and carries no current in a perfectly balanced three-phase arrangement under a balanced sinusoidal steady state operation. For unbalanced arrangements, including the effect of the ground plane and the shield wires, computer programs are available that can also include frequency dependence of the transmission-line parameters.

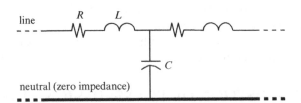

FIGURE 4.3 Distributed parameter representation on a per-phase basis.

4.4.1 Resistance R

The resistance of a transmission line per-unit is designed to be small to minimize I^2R power losses. These losses go down as the conductor size increases, but the costs of conductors and towers increase. In the United States, the Department of Energy estimates that approximately 9% of generated electricity is lost in transmission and distribution. Therefore, it is important to keep the transmission-line resistance small. In a bundled arrangement, the parallel resistance of the bundled conductors is under consideration.

There are tables [2] that give the resistance at DC and 60 Hz frequency at various temperatures. ACSR conductors such as that shown in Figure 4.1c are considered hollow, as shown in Figure 4.4a, because the electrical resistance of the steel core is much higher than that of the outer aluminum strands.

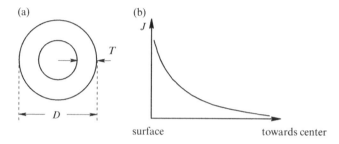

FIGURE 4.4 (a) Cross-section of an aluminum conductor steel reinforced conductor. (b) Skin effect in a solid conductor.

Also, the skin-effect phenomenon plays a role at 60 Hz (or 50 Hz) frequency. The line resistance R depends on the length of the conductor l, the resistivity of the material ρ (which increases with temperature), and (inversely) the effective cross-sectional area A of the conductor through which the current flows:

$$R = \frac{\rho l}{A} \tag{4.1}$$

The effective area A in Equation 4.1 depends on the frequency due to the skin effect, where the current at 60 Hz frequency is not uniformly distributed throughout the cross-section; rather, it crowds toward the periphery of the conductor, with a higher current-density J, as shown in Figure 4.4b, for a *solid* conductor (it is more uniform in a hollow conductor), and decreases exponentially such that at the skin depth, the current density is a factor of $e (= 2.718)$ smaller than that at the surface. The skin depth of a material at a frequency f is [4]

$$\delta = \sqrt{\frac{2\rho}{(2\pi f)\mu}} \tag{4.2}$$

which is calculated for aluminum in the following example.

Example 4.1
In Figure 4.4a, the conductor is aluminum with resistivity $\rho = 2.65 \times 10^{-8}\,\Omega - m$. Calculate the skin depth δ at a 60 Hz frequency.

Solution The permeability of aluminum can be considered that of free space: that is, $\mu = 4\pi \times 10^{-7}$ H/m. The resistivity is $\rho = 2.65 \times 10^{-8} \Omega - m$. Substituting these values in Equation 4.2,

$$\delta = 10.58\,\text{mm}$$

Therefore, in ACSR conductors, the aluminum thickness T, as shown in Figure 4.4a, is kept on the order of the skin depth. Any further thickness of the aluminum is essentially a waste that will not result in decreasing the overall resistance to AC currents. In the case of a type of ACSR conductor called a Bunting conductor, shown in Figure 4.4a, $T / D = 0.3748$, where $D = 1.302$ in. Therefore, the skin effect results in resistance only slightly higher at 60 Hz compared to at DC in such a hollow conductor, where the resistance at DC is listed as 0.0787 ohms/mile versus 0.0811 ohms/mile at 60 Hz, both at atemperature of 25° C [2].

4.4.2 Shunt Conductance G

In transmission lines, in addition to the power loss due to $I^2 R$ in the series resistance, there is a small loss due to the leakage current flowing through the insulator. This effect is amplified due to the corona effect, where the surrounding air is ionized, and a hissing sound can be heard in misty, foggy weather. The corona problem can be averted by increasing the conductor size and using conductor bundling, as discussed earlier [2]. These losses can be represented by putting a conductance G in the shunt with the capacitance in Figure 4.3 since such losses depend approximately on the square of the voltage. However, these losses are negligibly small; thus, in our analysis, we neglect the presence of G.

4.4.3 Series Inductance L

Figure 4.5a Shows a balanced three-phase transmission line with currents i_a, i_b, and i_c, which add up to zero under balanced conditions. That is, at any time,

$$i_a + i_b + i_c = 0 \tag{4.3}$$

We will calculate the inductance of each phase, such as phase a, as the ratio of the total flux linking it to its current. The total flux linking a phase conductor by the three currents can be obtained by superposition, and thus the per-phase inductance can be calculated as

$$L_a = \frac{\lambda_{a,\text{total}}}{i_a} = \frac{1}{i_a}(\lambda_{a,i_a} + \lambda_{a,i_b} + \lambda_{a,i_c}) \tag{4.4}$$

Considering i_a by itself, as shown in Figure 4.5b, by Ampere's law at a distance x from conductor a,

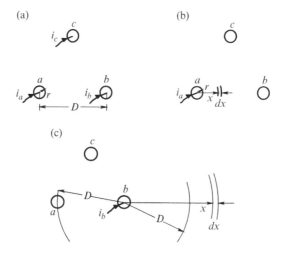

FIGURE 4.5 Flux linkage with conductor a, where r is the conductor radius. (a) Balanced three-phase transmission line with currents; (b) Ampere's law on conductor a; (c) mutual flux linking conductors a and b.

$$H_x = \frac{i_a}{2\pi x} \rightarrow \text{and} \rightarrow B_x = \mu_0 H_x = \left(\frac{\mu_0}{2\pi x}\right) i_a \qquad (4.5)$$

Therefore, in Figure 4.5b, the differential flux linkage in a differential distance dx over a unit length along the conductor ($\ell = 1$) is

$$d\lambda_{x,i_a} = B_x \cdot dx = \left(\frac{\mu_0}{2\pi x}\right) i_a \cdot dx \qquad (4.6)$$

Assuming the current in each conductor to be at the surface (a reasonable assumption, based on the discussion of the skin effect in calculating line resistances), integrating x from the conductor radius to infinity,

$$\lambda_{a,i_a} = \int_r^\infty d\lambda_{x,i_a} = \left(\frac{\mu_0}{2\pi}\right) i_a \int_r^\infty \frac{1}{x} \cdot dx = \left(\frac{\mu_0}{2\pi}\right) i_a \ln\frac{\infty}{r} \qquad (4.7)$$

Next, we will calculate the mutual flux linking conductor a due to i_b, as shown in Figure 4.5c. Noting that $D \gg r$, the flux linking conductor a due to i_b is between a distance of D and infinity. Therefore, using the previous procedure and Equation 4.7,

$$\lambda_{a,i_b} = \left(\frac{\mu_0}{2\pi}\right) i_b \ln\frac{\infty}{D} \qquad (4.8)$$

Similarly, due to i_c,

$$\lambda_{a,i_c} = \left(\frac{\mu_0}{2\pi}\right) i_c \ln\frac{\infty}{D} \qquad (4.9)$$

Therefore, adding the flux-linkage components from Equations 4.7 through 4.9,

$$\lambda_{a,\text{total}} = \lambda_{a,i_a} + \lambda_{a,i_b} + \lambda_{a,i_c} = \left(\frac{\mu_0}{2\pi}\right)\left[i_a \ln\frac{\infty}{r} + (i_b + i_c)\ln\frac{\infty}{D}\right] \qquad (4.10)$$

From Equation 4.3, the sum of i_b and i_c equals $(-i_a)$ in Equation 4.10, which simplifies to

$$\lambda_{a,\text{total}} = \left(\frac{\mu_0}{2\pi}\right) i_a \ln\frac{D}{r} \qquad (4.11)$$

Therefore, from Equation 4.4, the inductance L associated with each phase per-unit length is

$$L = \left(\frac{\mu_0}{2\pi}\right)\ln\frac{D}{r} \rightarrow (\text{H/m}) \qquad (4.12)$$

If the three conductors are at distances D_1, D_2, and D_3 with respect to each other, as shown in Figure 4.2 but transposed, then the equivalent distance D (also known as the geometric mean distance [GMD]) between them, for use in Equation 4.12, can be calculated as

$$D = \sqrt[3]{D_1 D_2 D_3} \qquad (4.13)$$

In these calculations, it is assumed that the current of each conductor flows entirely at the surface. However, various computer programs can consider the internal conductor inductance based on the current-density distribution internal to the conductor at a given frequency.

In bundled conductors, the inductance per-unit length is smaller: for example, by a factor of 0.7 if a three-conductor bundle is used with a spacing of 18 in., as shown in Figure 4.1b [2]. This factor is approximately 0.8 in a two-conductor bundle with a spacing of 18 in.

4.4.4 Shunt Capacitance *C*

To calculate line capacitances, consider a charge q on a conductor, as shown in Figure 4.6, that results in dielectric flux lines and the electric field intensity E. Consider a surface at a distance x from the conductor of unit length along the conductor. The surface area per-unit length, perpendicular to the field lines, is $(2\pi x)\times 1$. Therefore, the dielectric flux density D and the electric field E can be calculated as

$$D = \frac{q}{(2\pi x) \times 1} \rightarrow \text{and} \rightarrow E = \frac{D}{\varepsilon_0} = \frac{q}{(2\pi x)\varepsilon_0} \tag{4.14}$$

where $\varepsilon_0 = 8.85 \times 10^{-12}\,\text{F/m}$ is the permittivity of free space and approximately the same for air.

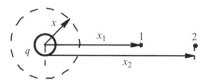

FIGURE 4.6 Electric field due to a charge.

Therefore, in Figure 4.6, the voltage of point 1 with respect to point 2, v_{12}, is

$$v_{12} = -\int_{x_2}^{x_1} E(x)\cdot dx = \left(\frac{q}{2\pi\varepsilon_0}\right)\ln\frac{x_2}{x_1} \tag{4.15}$$

Considering three conductors at symmetrical distances D (not to be confused with the dielectric flux density), as shown in Figure 4.7a, we can calculate voltage v_{ab}, for example, by superposing the effects of q_a, q_b, and q_c, as follows.

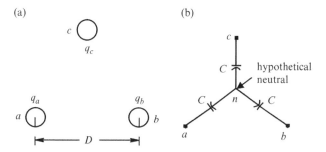

FIGURE 4.7 Shunt capacitances. (a) Three conductors at symmetrical distances; (b) capacitances from each phase.

From Equation 4.15, where r is the conductor radius, due to q_a,

$$v_{ab,q_a} = \left(\frac{q_a}{2\pi\varepsilon_0}\right)\ln\frac{D}{r} \tag{4.16}$$

Similarly, due to q_b,

$$v_{ba,q_b} = \left(\frac{q_b}{2\pi\varepsilon_0}\right)\ln\frac{D}{r} = -v_{ab,q_b} \tag{4.17}$$

where $v_{ab,q_b} = -\left(\dfrac{q_b}{2\pi\varepsilon_0}\right)\ln\dfrac{D}{r}$.

Note that q_c, being equidistant from both conductors a and b, does not produce any voltage between a and b: that is, $v_{ab,q_c} = 0$. Therefore, superposing the effects of all three charges,

$$v_{ab} = v_{ab,q_a} + v_{ab,q_b} + \underbrace{v_{ab,q_c}}_{(=0)} = \frac{1}{2\pi\varepsilon_0}(q_a - q_b)\ln\frac{D}{r} \tag{4.18}$$

Now consider a hypothetical neutral point n, as shown in Figure 4.7b, and the capacitances C from each phase as indicated, where

$$v_{ab} = v_{an} - v_{bn} = \frac{q_a}{C} - \frac{q_b}{C} = \frac{1}{C}(q_a - q_b) \tag{4.19}$$

Comparing Equations 4.18 and 4.19, the shunt capacitance per-unit length is

$$C = \frac{2\pi\varepsilon_0}{\ln\dfrac{D}{r}} \rightarrow (\text{F/m}) \tag{4.20}$$

If the three conductors are at distances D_1, D_2, and D_3, as shown in Figure 4.2, but transposed, then the equivalent distance (GMD) D between them, for use in Equation 4.20, can be calculated as

$$D = \sqrt[3]{D_1 D_2 D_3} \tag{4.21}$$

In Equation 4.20, it should be noted that the per-phase capacitance does not include the effect of ground and the presence of shield wires. This effect is included in [5].

In bundled conductors, the per-unit shunt capacitance is larger by a factor of approximately 1.4 if a three-conductor bundle is used with a spacing of 18 in. [2]. This factor is approximately 1.25 in a two-conductor bundle with a spacing of 18 in.

Typical values of transmission-line parameters at various voltage levels are given in Table 4.1, where it is assumed that the conductors are arranged in three-conductor bundles with a spacing of 18 in. [6].

TABLE 4.1 Approximate transmission-line parameters with bundled conductors at 60 Hz

Nominal voltage	$R(\Omega/\text{km})$	$\omega L(\Omega/\text{km})$	$\omega C(\mu\text{U}/\text{km})$
230 kV	0.06	0.5	3.4
345 kV	0.037	0.376	4.518
500 kV	0.03	0.33	5.3
765 kV	0.01	0.34	5.0

Example 4.2

Consider a 345 kV transmission line of 3L3 type towers, shown in Figure 4.8. This transmission system consists of a single conductor per phase, which is a Bluebird ACSR conductor with a diameter of 1.762 in. Ignoring the effects of ground, ground wires, and conductor sags, calculate $\omega L(\Omega/\text{km})$ and $\omega C(\mu\text{U}/\text{km})$, and compare the results with those given in Table 4.1.

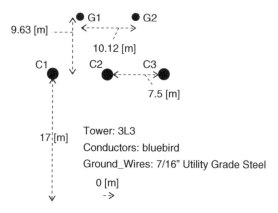

FIGURE 4.8 A 345 kV, single-conductor per-phase transmission system.

Solution The distances between the conductors in Figure 4.8 are 7.5 m, 7.5 m, and 15 m, respectively. Therefore, from Equation 4.21, the equivalent distance between them is $D = \sqrt[3]{7.5\times7.5\times15} = 9.45\,\text{m}$, and the radius is $r = 0.0224\,\text{m}$. Therefore at 60 Hz, from Equations 4.12 and 4.20, $\omega L = 0.456\,\Omega/\text{km}$ and $\omega C = 3.467\,\mu\text{U}/\text{km}$. Accurate values, including ground wires, were calculated using PSS®E, and the corresponding program file and results are on the associated website. These results, as follows, are very close to the analytical calculations: $\omega L = 0.474\ \Omega/\text{km}$ and $\omega C = 3.523\,\mu\text{U}/\text{km}$.

4.5 DISTRIBUTED-PARAMETER REPRESENTATION OF TRANSMISSION LINES IN A SINUSOIDAL STEADY STATE

A three-phase transmission line, assumed to be perfectly transposed, can be treated on a per-phase basis, as shown in Figure 4.9, under balanced operation in a sinusoidal steady state.

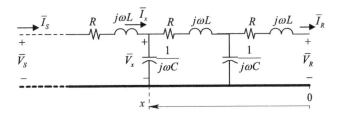

FIGURE 4.9   Distributed per-phase transmission line (G not shown).

Transmission lines are designed to have a resistance as small as economically feasible. Therefore, in transmission lines of medium length (300 km or so) or shorter, it is reasonable to assume their resistance is lumped, which can be taken into account separately.

With the distance x measured from the receiving end, as shown in Figure 4.9, and omitting the resistance R for now, at a distance x, for a small distance Δx,

$$\Delta \bar{V}_x = j\omega(L\Delta x)\bar{I}_x \rightarrow \text{or} \rightarrow \frac{d\bar{V}_x}{dx} = j\omega L\bar{I}_x \tag{4.22}$$

Similarly, the current flowing through the distributed shunt capacitance, for a small distance Δx, is

$$\Delta \bar{I}_x = j\omega(C\Delta x)\bar{V}_x \rightarrow \text{or} \rightarrow \frac{d\bar{I}_x}{dx} = j\omega C\bar{V}_x \tag{4.23}$$

Differentiating Equation 4.22 with respect to x and substituting into Equation 4.23,

$$\frac{d^2\bar{V}_x}{dx^2} + \beta^2\bar{V}_x = 0 \tag{4.24}$$

where

$$\beta = \omega\sqrt{LC} \tag{4.25}$$

is the propagation constant. Similarly, for the current,

$$\frac{d^2 \bar{I}_x}{dx^2} + \beta^2 \bar{I}_x = 0 \tag{4.26}$$

We should note that under the assumption of perfect transposition that allows us to treat a three-phase transmission line on a per-phase basis, the voltage equation in Equation 4.24 is independent of current, while the current equation in Equation 4.26 is independent of voltage. The solution of Equation 4.24 has the following form:

$$\bar{V}_x(s) = \bar{V}_1 e^{\beta jx} + \bar{V}_2 e^{-j\beta x} \tag{4.27}$$

where $\bar{V}_1$ and $\bar{V}_2$ are coefficients calculated based on the boundary conditions. The current in Equation 4.26 has a solution with a form similar to Equation 4.27. Differentiating Equation 4.27 with respect to x,

$$\frac{d\bar{V}_x}{dx} = j\beta(\bar{V}_1 e^{j\beta x} - \bar{V}_2 e^{-j\beta x}) \tag{4.28}$$

From Equations 4.28 and 4.22,

$$\bar{I}_x = (\bar{V}_1 e^{j\beta x} - \bar{V}_2 e^{-j\beta x}) / Z_c \tag{4.29}$$

where Z_c is the surge impedance

$$Z_c = \sqrt{\frac{L}{C}} \tag{4.30}$$

Applying the boundary conditions in Equations 4.27 and 4.29 – that is, at the receiving end – $x = 0$, $\bar{V}_{x=0} = \bar{V}_R$, and $\bar{I}_{x=0} = \bar{I}_R$,

$$\bar{V}_1 + \bar{V}_2 = \bar{V}_R \rightarrow \text{and} \rightarrow \bar{V}_1 - \bar{V}_2 = Z_c \bar{I}_R \tag{4.31}$$

From Equation 4.31,

$$\bar{V}_1 = \frac{\bar{V}_R + Z_c \bar{I}_R}{2} \rightarrow \text{and} \rightarrow \bar{V}_2 = \frac{\bar{V}_R - Z_c \bar{I}_R}{2} \tag{4.32}$$

Substituting from Equation 4.32 into Equation 4.27, and recognizing $\frac{e^{j\beta x} + e^{-j\beta x}}{2} = \cos \beta x$ and $\frac{e^{-j\beta x} - e^{-j\beta x}}{2} = j \sin \beta x$,

$$\bar{V}_x = \bar{V}_R \cos \beta x + j Z_c \bar{I}_R \sin \beta x \tag{4.33}$$

Similarly, substituting from Equation 4.32 into Equation 4.29,

$$\bar{I}_x = \bar{I}_R \cos \beta x + j \frac{\bar{V}_R}{Z_c} \sin \beta x \qquad (4.34)$$

4.6 SURGE IMPEDANCE Z_C AND SURGE IMPEDANCE LOADING (SIL)

Consider that a lossless transmission line is loaded by a resistance that equals Z_c, as shown in Figure 4.10a. Assuming the receiving-end voltage to be the reference phasor,

$$\bar{V}_R = V_R \angle 0° \rightarrow \text{and} \rightarrow \bar{I}_R = \frac{V_R}{Z_c} \angle 0° \qquad (4.35)$$

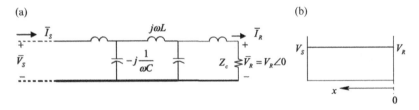

(a)   (b)

FIGURE 4.10 Per-phase transmission line terminated with a resistance equal to Z_c. (a) Lossless transmission line loaded with a resistance Z_c; (b) voltage profile.

Therefore, from Equations 4.33 and 4.35, recognizing that $\bar{V}_R = V_R \angle 0°$,

$$\bar{V}_x = V_R(\cos \beta x + j \sin \beta x) = V_R e^{j\beta x} \qquad (4.36)$$

Similarly,

$$\bar{I}_x = I_R(\cos \beta x + j \sin \beta x) = I_R e^{j\beta x} \qquad (4.37)$$

The voltage equation given by Equation 4.36 shows that the voltage magnitude has a flat profile, as shown in Figure 4.10b: that is, it is the same magnitude everywhere on the line; only the angle increases with distance x. This loading is called the *surge impedance loading* (SIL); the reactive power consumed by the line everywhere is the same as the reactive power produced

$$\omega L I_x^2 = V_x^2 \omega C \qquad (4.38)$$

which is true, since under SIL, from Equations 4.36 and 4.37, $\bar{V}_x / \bar{I}_x = Z_c$. For a given voltage level, the characteristic impedance of a transmission line falls in a narrow range because the separation and height of the conductors above ground depend on this voltage level. Table 4.2 shows the typical values for the surge impedance and three-phase SIL, where

$$\text{SIL} = \frac{V_{LL}^2}{Z_c} \tag{4.39}$$

TABLE 4.2 Approximate surge impedance and three-phase surge impedance loading

Nominal voltage	$Z_c(\Omega)$	SIL (MW)
230 kV	385	135 MW
345 kV	275	430 MW
500 kV	245	1020 MW
765 kV	255	2300 MW

From Table 4.2, it is clear that a large transfer of power, such as 1,000 MW, calls for a high-voltage line, such as 345 kV or 500 kV; otherwise, many circuits in parallel will be needed.

Example 4.3
For the 345 kV transmission line described in Example 4.2, calculate the surge impedance Z_c and the SIL. If the line is 100 km long and the line resistance is 0.031Ω/km , calculate the percentage loss if the line is surge-impedance loaded.

Solution   In the single-conductor transmission line from Example 4.2, we calculated that $\omega L = 0.456\Omega$/km and $\omega C = 3.467\,\mu\text{℧}$/km. Therefore, from Equation 4.30, $Z_c \approx 363\Omega$; and from Equation 4.39, SIL $= 328\,\text{MW}$. At SIL of this transmission line, the per-phase current through the line is $I = \dfrac{345\times10^3 / \sqrt{3}}{Z_c} = 548.7\,\text{A}$, and therefore, the power loss as a percentage of SIL in this 100 km line is

$$\% I^2 R = 100 \times \frac{3\times 548.7^2 \times (0.031\times 100)}{328\times 10^6} \approx 0.85\% \rightarrow\rightarrow\rightarrow$$

4.6.1   Line Loadability

The SIL provides a benchmark in terms of which the maximum loading of a transmission line can be expressed [2, 6, 7]. This loading is a function of the length of the transmission line so that certain constraints are met. This approximate loadability is a function of the line length. Short lines less than 100 km can be loaded more than 3 times SIL, based on not exceeding the thermal limits. Medium-length lines (100–300 km) can be loaded to 1.5 to 3 times the SIL before the voltage drop across them exceeds 5%. Long lines (more than 300 km) can be loaded to around the SIL because of the stability limit (discussed later in this book) so that the phase angle of the voltage between the two ends does not exceed 40–45 degrees. The loadability of medium and long lines can be increased by providing series and shunt compensation, as discussed in Chapter 10. These results are summarized in Table 4.3.

TABLE 4.3   Approximate loadability of transmission lines

Line length (km)	Limiting factor	Multiple of SIL
< 100	Thermal	> 3
100–300	5% voltage drop	1.5–3
> 300	Stability	1.0–1.5

4.7   LUMPED TRANSMISSION-LINE MODELS IN A STEADY STATE

In a balanced sinusoidal steady state, it is very useful to have a per-phase model of transmission lines on the assumption that the three phases are perfectly transposed and the applied voltages and currents are balanced and in a sinusoidal steady state.

In the distributed-parameter representation of the transmission line in the previous section, at the sending end with $x = \ell$,

$$\bar{V}_S = \bar{V}_R \cos \beta \ell + j Z_c \bar{I}_R \sin \beta \ell \qquad (4.40)$$

and

$$\bar{I}_S = j \frac{\bar{V}_R}{2} \sin \beta \ell + \bar{I}_R \cos \beta \ell \qquad (4.41)$$

This transmission line can be represented by the two-port shown in Figure 4.11, where the symmetry is necessary from both ports due to the bilateral nature of the transmission line. To find Z_{series} and $Y_{\text{shunt}} / 2$, the following procedure is followed: hypothetically short-circuiting the receiving end results in $\bar{V}_R = 0$; and from Figure 4.11 and Equation 4.40,

$$\left.\frac{\bar{V}_S}{\bar{I}_R}\right|_{\bar{V}_R=0} = Z_{\text{series}} = jZ_c \sin \beta\ell \tag{4.42}$$

For transmission lines of medium length or shorter, $\beta\ell$ is small (see Problem 4.14). Recognizing that $\beta\ell$ is small, the following approximation holds: $\sin \beta\ell \approx \beta\ell$. Making use of this approximation in Equation 4.42, and recognizing that $\beta = \omega\sqrt{LC}$ and $Z_c = \sqrt{L/C}$,

$$Z_{\text{series}} = j\omega L_{\text{line}} \rightarrow \text{where} \rightarrow L_{\text{line}} = \ell L \tag{4.43}$$

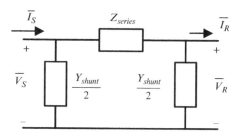

FIGURE 4.11 Lumped representation.

With the receiving end open-circuited in Figure 4.11, $\bar{I}_R = 0$, and from Equation 4.40,

$$\left.\frac{\bar{V}_S}{\bar{V}_R}\right|_{\bar{I}_R=0} = 1 + Z_{\text{series}}\frac{Y_{\text{shunt}}}{2} = \cos \beta\ell \tag{4.44}$$

Therefore, from Equations 4.43 and 4.44,

$$\left(\frac{Y_{\text{shunt}}}{2}\right) = \frac{\cos \beta\ell - 1}{j\omega L_{\text{line}}} \tag{4.45}$$

From the Taylor series expansion for small values of $\beta\ell$, $\cos \beta\ell \approx 1 - \dfrac{(\beta\ell)^2}{2}$; and therefore, in Equation 4.45,

$$\frac{Y_{\text{shunt}}}{2} = j\frac{\omega C_{\text{line}}}{2} \rightarrow \text{where} \rightarrow C_{\text{line}} = C\ell \tag{4.46}$$

Therefore, the per-phase equivalent circuit representation in Figure 4.11 becomes as shown in Figure 4.12, where the series resistance is explicitly shown as a separate element. This can be confirmed from the expressions derived in the appendix for long-length lines, where the line resistance R is also considered distributed and simplified for medium-length lines (see Problem 4.13).

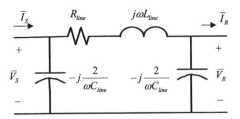

FIGURE 4.12   Per-phase representation for medium-length transmission lines.

4.7.1   Short-Length Lines

In short lines (less than 100 km), the effect of capacitive vars is small compared to the system strength to provide Q and can often be neglected. Therefore, the shunt capacitances shown in Figure 4.12 can be neglected. This results in only the series impedance, where even the series resistance may be neglected in some simplified studies.

4.7.2   Long Lines

As derived in Appendix A, for long lines exceeding 300 km, it is prudent to use a more exact representation. The parameters in the per-phase equivalent circuit from Figure 4.11 are as follows:

$$Z_{\text{series}} = Z_c \sinh \gamma \ell \tag{4.47}$$

$$\frac{Y_{\text{shunt}}}{2} = \frac{\tanh\left(\dfrac{\gamma \ell}{2}\right)}{Z_c} \tag{4.48}$$

where

$$Z_c = \sqrt{\frac{R + j\omega L}{G + j\omega C}} \rightarrow \text{and} \tag{4.49}$$

$$\gamma = \sqrt{(R + j\omega L)(G + j\omega C)} \tag{4.50}$$

4.8   CABLES

As stated in [8], the underground transmission cable usage in the United States is very small: less than 1% of overhead line mileage. The highest underground cable voltage commonly used in the United States is 345 kV, and a large portion of this cable system is high-pressure fluid-filled pipe-type cable. Extruded dielectric cables are commonly used in the United States for up to 230 kV, with up to 500 kV in service overseas. Underground transmission cable is generally more expensive than overhead lines. Because of all the variables (system design, route considerations, cable type, raceway type, etc.), it has to be determined case by case whether underground transmission cable is a viable alternative. A rule of thumb is that underground transmission cable will cost 3 to 20 times as much as overhead line construction. As a result of the high cost, the use of high-voltage power cable for transmission and subtransmission is generally limited to special applications caused by environmental and/or right-of-way restrictions. If underground transmission cable is to be considered, an engineering study is required to properly evaluate the possible underground alternatives.

Underground cables are used for power transfer in and around metropolitan areas due to a lack of space for overhead transmission lines or for aesthetic reasons. These cables are also more reliable and not exposed to the elements like overhead transmission lines. For long-distance bulk power transmission, underground cables generally are not used because of their cost compared to overhead transmission lines, although new trenching technologies and cable materials may reduce this cost disadvantage.

Underground cables have much larger capacitance than overhead transmission lines; hence, their characteristic impedance Z_c is much smaller. However, despite lower values of Z_c and higher values of SIL, the loading of cables is limited by the problem of getting rid of the dissipated heat. For underwater transmission, cables are used in DC systems for the following reason: because of the high capacitance of underground cables, operating at 60/50 Hz AC would require shunt reactors at periodic distances to compensate for the capacitive charging currents. Such underwater compensation reactors are not feasible.

There are various types of cables in use: pipe-type oil-filled cables, self-contained oil-filled cables with a single core, cross-linked cables with polyethylene insulation, and gas-insulated cables with compressed SF_6 gas [9]. Research is being done on superconducting cables as well. The modeling of cable parameters for power system studies is like that for overhead transmission lines.

REFERENCES

1. United States Department of Agriculture. *Design Manual for High Voltage Transmission Lines*. RUS Bulletin 1724E-200.
2. Electric Power Research Institute (EPRI). *Transmission Line Reference Book: 345 kV and Above*, 2nd ed.
3. H. W. Dommel. 1986. *EMTP Theory Book*. BPA.
4. N. Mohan, T. Undeland, and W. P. Robbins. 2003. *Power Electronics: Converters, Applications, and Design*, 3rd ed. John Wiley & Sons.

5. PSS®E high-performance transmission planning and analysis software. https://new.
   siemens.com/global/en/products/energy/energy-automation-and-smart-grid/pss-
   software/pss-e.html.
6. P. Kundur. 1994. *Power System Stability and Control*. McGraw Hill.
7. R. Dunlop et al. 1979. "Analytical Development of Loadability Characteristics for
   EHV and UHV Transmission Lines." *IEEE Transactions on PAS*. Vol. PAS-98, No.2.
8. United States Department of Agriculture. *Rural Utilities Service, Design Guide for
   Rural Substations*. RUS Bulletin 1724E-300.
9. EPRI. *Underground Transmission Systems Reference Book*.

PROBLEMS

4.1 According to [1], the conductor surface gradient in a single-conductor
per-phase transmission line can be calculated as follows

$$g = \frac{k V_{LL}}{\sqrt{3}\, r \ln \dfrac{D}{r}}$$

where $k V_{LL}$ is line-line voltage in kV, r is the conductor radius in cm,
D is the geometric mean distance (GMD) of the phase conductors in
cm, and g is the conductor surface gradient in kV/cm.

Calculate the conductor surface gradient for a 230 kV line (a) with
Dove ACSR conductors for which $r = 1.18$ cm, and (b) with Pheasant
ACSR conductors for which $r = 1.755$ cm. Assume that $D = 784.9$ cm in
both cases.

4.2 According to [1], the conductor surface gradient in a two-conductor
bundle per-phase transmission line can be calculated as follows

$$g = \frac{k V_{LL}(1 + 2r/s)}{2\sqrt{3}\, r \ln \dfrac{D}{\sqrt{rs}}}$$

where $k V_{LL}$ is the line-line voltage in kV, r is the conductor radius in cm,
D is the GMD of the phase conductors in cm, s is the separation between subconductors in cm, and g is the conductor surface gradient
in kV/cm.

Calculate the conductor surface gradient for a 345 kV line with a
two-conductor bundle per phase with Drake ACSR conductors for
which $r = 1.407$ cm. Assume that $D = 914$ cm and $s = 45.72$ cm (18 in.).

4.3 The parameters for a 500 kV transmission system shown in Figure 4.1
with bundled conductors are as follows: $Z_c = 258\,\Omega$ and
$R = 1.76 \times 10^{-2}\,\Omega/\text{km}$. Calculate the value for the surge impedance

loading (SIL) and the percentage power loss in this transmission line if it is 300 km long and is loaded to its SIL.

4.4 The 500 kV line of the type in Problem 4.3 is short: 80 km. It is loaded to three times its SIL. Calculate the power loss in this line as a percentage of its loading.

4.5 The 345 kV transmission line of the type described in Example 4.3 is 300 km long. Calculate the parameters for its model as a long line. Compare these parameter values for the model that assumes it to be a medium-length line. Compare the percentage errors in assuming it to be a medium-length line.

4.6 Consider a 345 kV transmission line with parameters similar to those described in Table 4.1. Assuming a 100 MVA base, obtain its parameters in per-unit.

4.7 Consider a 300 km, 345 kV transmission line with parameters similar to those described in Table 4.1. Assume its receiving-end voltage $V_R = 1.0$ pu. Plot the voltage ratio V_s / V_R as a function of P_R / SIL, where P_R is the unity power factor load at the receiving end. P_R / SIL varies in a range from 0 to 3.

4.8 Repeat Problem 4.7 if the load impedance has a power factor of (a) 0.9 lagging and (b) 0.9 leading. Compare these results with a unity-power-factor loading.

4.9 Calculate the phase angle difference of voltages between the two ends of the line in Problem 4.7 that has unity-power-factor loading.

4.10 A 200 km, 345 kV line has the parameters given in Table 4.1. Neglect the resistance. Calculate the voltage profile along this line if it is loaded to (a) 1.5 times SIL and (b) 0.75 times SIL if both ends are held at the voltage of 1 per-unit.

4.11 In Problem 4.10, calculate the reactive power at both ends under the two loading levels.

4.12 In a 230 kV cable system, the surge impedance (ignoring losses) is 25 ohms, and the charging is 14.5 MVA/km. The cable length is 20 km. The resistance is 0.05 ohms/km. Both ends of this cable system are held at 1 per-unit, with the sending-end voltage $\bar{V}_s = 1.0\angle 10°$ pu and the receiving-end voltage $\bar{V}_r = 1.0\angle 0°$ pu. Calculate the power loss in the cable, and express it as a percentage of the power received at the receiving end.

4.13 Show that the expressions derived in the appendix for long-length lines, where the line resistance R is also considered distributed, simplify to expressions for medium-length lines that lead to Figure 4.12.

4.14 Show that for a 345 kV, 200 km, medium-length line, $\beta\ell$ is small such that the approximations $\sin\beta\ell \approx \beta\ell$ and $\cos\beta\ell \approx 1 - \dfrac{(\beta\ell)^2}{2}$ are reasonable.

PSS®E-BASED PROBLEMS

4.15   Calculate the parameters for the single-conductor per-phase 345 kV transmission line in Example 4.2, using the data file for this problem on the accompanying website.

4.16   Calculate the parameters for the two-conductor-bundle per-phase 345 kV transmission line using the data file for this problem on the accompanying website.

4.17   Calculate the parameters for the three-conductor-bundle per-phase 500 kV transmission line using the data file for this problem on the accompanying website.

4.18   Obtain the results for Problem 4.11 in PSS®E using the data file for this problem on the accompanying website.

APPENDIX 4A   LONG TRANSMISSION LINES

In this appendix, we will consider transmission lines exceeding 300 km. The procedure for determining their representation is similar to that for the medium-length lines, except the line resistance R is also considered distributed, and the distributed shunt conductance G is included.

With the distance x measured from the receiving end, as shown in Figure 4.9,

$$\frac{d\bar{V}_x}{dx} = (j\omega L + R)\bar{I}_x \qquad (A4.1)$$

Similarly, due to the current flowing through distributed shunt capacitance,

$$\frac{d\bar{I}_x}{dx} = (j\omega C + G)\bar{V}_x \qquad (A4.2)$$

Differentiating Equation A4.1 with respect to x and substituting into Equation A4.2,

$$\frac{d^2\bar{V}_x}{dx^2} = (j\omega L + R)(j\omega C + G)\bar{V}_x = \gamma^2 \bar{V}_x \qquad (A4.3)$$

where

$$\gamma = \sqrt{(R + j\omega L)(G + j\omega C)} = \alpha + j\beta \qquad (A4.4)$$

is the propagation constant where α and β are positive in values. Similarly, for the current,

$$\frac{d^2\bar{I}_x}{dx^2} = (R + j\omega L)(G + j\omega C)\bar{I}_x = \gamma^2 \bar{I}_x \qquad (A4.5)$$

The solution of Equation A4.3 is of the following form

$$\bar{V}_x = \bar{V}_1 e^{\gamma x} + \bar{V}_2 e^{-\gamma x} \tag{A4.6}$$

where $\bar{V}_1$ and $\bar{V}_2$ are coefficients calculated based on the boundary conditions. The current in Equation A4.5 will have a solution with a form similar to Equation A4.6. Differentiating Equation A4.6 with respect to x,

$$\frac{d\bar{V}_x}{dx} = \gamma(\bar{V}_1 e^{\gamma x} - \bar{V}_2 e^{-\gamma x}) \tag{A4.7}$$

Comparing Equation A4.7 with Equation A4.1,

$$\bar{I}_x = (\bar{V}_1 e^{\gamma x} - \bar{V}_2 e^{-\gamma x}) / Z_c \tag{A4.8}$$

where Z_c is the surge impedance

$$Z_c = \sqrt{\frac{R + j\omega L}{G + j\omega C}} \tag{A4.9}$$

Applying the boundary conditions in Equations A4.6 and A4.8 – that is, at the receiving end – $x = 0$, $\bar{V}_{x=0} = \bar{V}_R$, and $\bar{I}_{x=0} = \bar{I}_R$:

$$\bar{V}_1 + \bar{V}_2 = \bar{V}_R \rightarrow \text{and} \rightarrow \bar{V}_1 - \bar{V}_2 = Z_c \bar{I}_R \tag{A4.10}$$

From Equation A4.10,

$$\bar{V}_1 = \frac{\bar{V}_R + Z_c \bar{I}_R}{2} \rightarrow \text{and} \rightarrow \bar{V}_2 = \frac{\bar{V}_R - Z_c \bar{I}_R}{2} \tag{A4.11}$$

Substituting from Equation A4.11 and recognizing $\dfrac{e^{\gamma x} + e^{-\gamma x}}{2} = \cosh \gamma x$ and $\dfrac{e^{\gamma x} - e^{-\gamma x}}{2} = \sinh \gamma x$,

$$\bar{V}_x = \bar{V}_R \cosh \gamma x + Z_c \bar{I}_R \sinh \gamma x \tag{A4.12}$$

Similarly,

$$\bar{I}_x = \frac{\bar{V}_R}{Z_c} \sinh \gamma x + \bar{I}_R \cosh \gamma x \tag{A4.13}$$

4A.1   LUMPED TRANSMISSION-LINE MODEL IN A STEADY STATE

In the distributed-parameter representation of the transmission line in the previous section at the sending end with $x = \ell$,

$$\bar{V}_S = \bar{V}_R \cosh \gamma\ell + Z_c \bar{I}_R \sinh \gamma\ell \qquad (A4.14)$$

and

$$\bar{I}_S = \frac{\bar{V}_R}{2} \sinh \gamma\ell + \bar{I}_R \cosh \gamma\ell \qquad (A4.15)$$

This transmission line can be represented by a two-port, shown in Figure 4.11, where the symmetry is necessary from both ports due to the bilateral nature of the transmission line. To find Z_{series} and $Y_{\text{shunt}} / 2$, the following procedure is followed: short-circuiting the receiving end results in $\bar{V}_R = 0$, and from Figure 4.11,

$$\left.\frac{\bar{V}_S}{\bar{I}_R}\right|_{\bar{V}_R=0} = Z_{\text{series}} = Z_c \sinh \gamma\ell \qquad (A4.16)$$

With the receiving end open-circuited in Figure 4.11, $\bar{I}_R = 0$,

$$\left.\frac{\bar{V}_S}{\bar{V}_R}\right|_{\bar{I}_R=0} = 1 + Z_{\text{series}} \frac{Y_{\text{shunt}}}{2} = \cosh \gamma\ell \qquad (A4.17)$$

Therefore,

$$\frac{Y_{\text{shunt}}}{2} = \frac{1}{Z_c}\left(\frac{\cosh \gamma\ell - 1}{\sinh \gamma\ell}\right) \qquad (A4.18)$$

From mathematical tables, for some arbitrary parameter A,

$$\frac{\cosh A - 1}{\sinh A} = \tanh\left(\frac{A}{2}\right) \qquad (A4.19)$$

and therefore

$$\frac{Y_{\text{shunt}}}{2} = \frac{\tanh\left(\dfrac{\gamma\ell}{2}\right)}{Z_c} \qquad (A4.20)$$

5

POWER FLOW IN POWER SYSTEM NETWORKS

5.1 INTRODUCTION

For planning purposes, it is important to know the power transfer capability of transmission lines to meet the anticipated load demand. It is also important to know the levels of power flow through various transmission lines under normal and certain contingency outage conditions to maintain the continuity of service. Knowledge of power flows and voltage levels under normal operating conditions is also necessary to determine fault currents if a line were to short-circuit, for example, and the ensuing consequences for the transient stability of the system.

All power companies commonly use a power-flow program that determines this power flow for planning and operation purposes. These power-flow calculations are usually performed on the bulk power system, where the effect of the underlying secondary network is implicitly included.

Determining power flow requires measurements of certain power system conditions. Theoretically, suppose all the bus voltages in terms of their magnitudes and phase angles could be measured confidently. In that case, the power-flow calculations could be carried out by solving the linear circuit in which voltages and branch impedances, including the load impedances, are all given. There would be no need for the procedure outlined in this chapter.

However, utilities measure a combination of quantities such as voltage magnitude V, real power P, and the reactive power Q at various buses. (Many fault-protection relays make these measurements anyway to perform their function, and the information gathered by them is utilized for

Electric Power Systems with Renewables: Simulations Using PSS®E, Second Edition. Ned Mohan and Swaroop Guggilam.
© 2023 John Wiley & Sons, Inc. Published 2023 by John Wiley & Sons, Inc.
Companion Website: www.wiley.com/go/mohaneps

power-flow calculations.) The measured information is telemetered to a central operating station through dedicated telephone, microwave or fiber optic links. This information has to be instantaneous – that is, measured simultaneously in time. The system to acquire these measurements and tele-meter them to a control center is called *supervisory control and data acquisition* (SCADA). It is recognized that measurement transducers do not always provide accurate information. To overcome this problem, the system is over-measured by measuring more quantities than are required. Then, a state estimator is used to throw away the "bad" and redundant measurements in a probabilistic manner.

In this chapter, we will study the basic formulation of the power-flow problem and discuss the most commonly used numerical solution methods to carry it out. To simplify our discussion, we will assume a balanced three-phase system; therefore, only a per-phase representation is necessary.

Next, we will describe how the system is represented for this study. The transmission lines are represented by their pi-circuit model with a series impedance Z_{series} ($= R + j\omega L$) and a susceptance $B_{shunt}/2$ ($= \omega C/2$) at each end of the line. These are expressed in per-unit at a common MVA and kV base. It is customary to use a three-phase 100 MVA base. Similarly, transformers are represented by their total leakage impedance expressed in per-unit in terms of the common MVA base, ignoring their magnetizing currents. Transformers are assumed to be at their nominal turns ratio; hence the turns ratios do not enter into per-unit-based calculations. Loads can be represented in a combination of different ways: by specifying their real and reactive powers (P and Q) as a constant-current source or by their impedance, which is treated as constant.

5.2   DESCRIPTION OF THE POWER SYSTEM

A power system can be considered to consist of the following buses, which are interconnected through transmission lines:

1. Load buses, where P and Q are specified. These are called PQ buses.
2. Generator buses, where the voltage magnitude V and the power P are speci-fied. These are called PV buses. If the upper and/or lower limits on the reac-tive power Q on a PV bus are specified, and this limit is reached, then such a bus is treated as a PQ bus where the reactive power is specified at the limit that is reached.
3. A slack bus is essentially an "infinite" bus, where the voltage magnitude V is specified (normally 1 pu), and its phase angle is assumed to be zero as a refer-ence angle. At this bus, P can be what it needs to be, based on the line losses, and hence it is called the slack bus, which takes up the slack. Similarly, Q at this bus can be what it needs to be to hold the voltage at the specified value.

4. There are buses where no P and Q injections are specified and the voltage is also not specified. Often, these become necessary for including transformers. These can be considered a subset of PQ buses with specified injections of $P = 0$ and $Q = 0$.
5. In practical applications, there are scenarios where the generators control the remote bus's voltage, introducing some new notions of bus types. This concept is discussed as part of Appendix 5B at the end of this chapter to keep the general discussion simple.

5.3 EXAMPLE POWER SYSTEM

To illustrate power-flow calculations, an extremely simple power system consisting of three buses is shown in Figure 5.1. These three buses are connected through three 345 kV transmission lines 200 km, 150 km, and 150 km long. Similar to the values listed in Table 4.1 in Chapter 4, assume that these transmission lines with bundled conductors have a series reactance of 0.376 Ω/km at 60 Hz and a series resistance of 0.037 Ω/km. The shunt susceptance $B (= \omega C)$ is 4.5 $\mu\mho$/km.

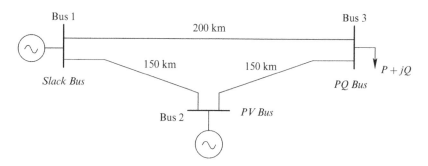

FIGURE 5.1 A three-bus 345 kV example system.

To convert the actual quantities to per-unit, the voltage base is 345 kV (L-L). Following the convention, a common three-phase 100 MVA base is chosen. Therefore, the base impedance is

$$Z_{base} = \frac{kV_{base}^2 (\text{phase})}{MVA_{base}(1-\phi)} = \frac{kV_{base}^2 (\text{L-L})}{MVA_{base}(3-\phi)} = 1190.25\Omega \qquad (5.1)$$

The base admittance $Y_{base} = 1/Z_{base}$. Based on the base impedance, base admittance, and line lengths, the line parameters and their per-unit values are given in Table 5.1.

TABLE 5.1 Per-unit values in the example system

Line	Series impedance Z in Ω (pu)	Total susceptance B in $\mu\mho$ (pu)
1–2	$Z_{12} = (5.55 + j56.4)\ \Omega = (0.0047 + j0.0474)$ pu	$B_{Total} = 675\mu\mho = (0.8034)$ pu
1–3	$Z_{13} = (7.40 + j75.2)\ \Omega = (0.0062 + j0.0632)$ pu	$B_{Total} = 900\mu\mho = (1.0712)$ pu
2–3	$Z_{23} = (5.55 + j56.4)\ \Omega = (0.0047 + j0.0474)$ pu	$B_{Total} = 675\mu\mho = (0.8034)$ pu

5.4 BUILDING THE ADMITTANCE MATRIX

It is easier to carry out the network solution in a nodal form than writing loop equations, which are more numerous. In writing the nodal equations, the current $\bar{I}_k$ is the current *injected* into a bus k, as shown in Figure 5.2 for the example system in Figure 5.1 [1–3].

By Kirchhoff's current law, the current injection at bus k is related to the bus voltages as follows

$$\bar{I}_k = \bar{V}_k Y_{kG} + \sum_{\substack{m \\ m \neq k}} \frac{\bar{V}_k - \bar{V}_m}{Z_{km}} \tag{5.2}$$

where Y_{kG} is the sum of the admittances connected at bus k to ground, and the second term consists of flows on all the lines connected to bus k, with the series impedance being Z_{km}, for example, between buses k and m. Therefore,

$$\bar{I}_k = \bar{V}_k \left(Y_{kG} + \sum_{\substack{m \\ m \neq k}} \frac{1}{Z_{km}} \right) - \sum_{\substack{m \\ m \neq k}} \frac{\bar{V}_m}{Z_{km}} \tag{5.3}$$

In Equation 5.3, the quantities within the brackets on the right side are designated as

$$Y_{kk} = Y_{kG} + \sum_{\substack{m \\ m \neq k}} \frac{1}{Z_{km}} \tag{5.4}$$

which is the self-admittance and is the sum of the admittances connected between bus k and the other buses, including ground. Similarly, in Equation 5.3, between buses k and m, the mutual admittance is

$$Y_{km} = -\frac{1}{Z_{km}} \tag{5.5}$$

which is negative of the inverse of the series impedance between buses k and m. This procedure allows the formulation of the bus-admittance matrix $[Y]$ for an n-bus system as

$$
\begin{bmatrix} \bar{I}_1 \\ \bar{I}_2 \\ \cdot\cdot \\ \cdot\cdot \\ \cdot\cdot \\ \bar{I}_n \end{bmatrix} =
\begin{bmatrix}
Y_{11} & Y_{12} & \cdot\cdot & \cdot\cdot & \cdot\cdot & Y_{1n} \\
Y_{21} & Y_{22} & \cdot\cdot & \cdot\cdot & \cdot\cdot & Y_{2n} \\
\cdot\cdot & \cdot\cdot & \cdot\cdot & \cdot\cdot & \cdot\cdot & \cdot\cdot \\
\cdot\cdot & \cdot\cdot & \cdot\cdot & \cdot\cdot & \cdot\cdot & \cdot\cdot \\
\cdot\cdot & \cdot\cdot & \cdot\cdot & \cdot\cdot & \cdot\cdot & \cdot\cdot \\
Y_{n1} & Y_{n2} & \cdot\cdot & \cdot\cdot & \cdot\cdot & Y_{nn}
\end{bmatrix}
\begin{bmatrix} \bar{V}_1 \\ \bar{V}_2 \\ \cdot\cdot \\ \cdot\cdot \\ \cdot\cdot \\ \bar{V}_n \end{bmatrix}
\tag{5.6}
$$

Example 5.1
In the example system in Figure 5.2, ignore all the shunt susceptances, and assemble the bus-admittance matrix of the form in Equation 5.6.

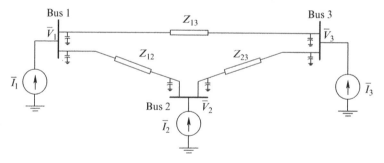

FIGURE 5.2 Example system from Figure 5.1 for assembling a Y-bus matrix.

Solution   In the example system, in per-unit, from Table 5.1,

$$Z_{12} = (0.0047 + j0.0474) \text{ pu}, Z_{13} = (0.0062 + j0.0632) \text{ pu},$$
$$Z_{23} = (0.0047 + j0.0474) \text{pu}$$

Therefore,

$$
\begin{bmatrix} \bar{I}_1 \\ \bar{I}_2 \\ \bar{I}_3 \end{bmatrix} =
\begin{bmatrix}
3.6090 - j36.5636 & -2.0715 + j20.8916 & -1.5374 + j15.6720 \\
-2.0715 + j20.8916 & 4.1431 - j41.7833 & -2.0715 + j20.8916 \\
-1.5374 + j15.6720 & -2.0715 + j20.8916 & 3.6090 - j36.5636
\end{bmatrix}
\begin{bmatrix} \bar{V}_1 \\ \bar{V}_2 \\ \bar{V}_3 \end{bmatrix}
$$

5.5   BASIC POWER-FLOW EQUATIONS

In the network equation in Equation 5.6, the current injections into buses are not specified explicitly. Rather, the injected real power P is specified at PV buses, and P and Q injections are specified at PQ buses as

$$P_k + jQ_k = \bar{V}_k \bar{I}_k^* \tag{5.7}$$

From the nodal equation in Equation 5.6,

$$I_k = \sum_{m=1}^{n} Y_{km} \bar{V}_m \text{ where } (Y_{km} = G_{km} + jB_{km}) \tag{5.8}$$

From Equation 5.8, recognizing that in terms of complex variables, if $a = bc$, then $a^* = b^* c^*$,

$$\bar{I}_k^* = \sum_{m=1}^{n} Y_{km}^* \bar{V}_m^* = \sum_{m=1}^{n} (G_{km} - jB_{km}) \bar{V}_m^* \tag{5.9}$$

Substituting Equation 5.9 into Equation 5.7,

$$P_k + jQk = \bar{V}_k \bar{I}_k^* = \sum_{m=1}^{n} \left[(G_{km} - jB_{km})(\bar{V}_k \bar{V}_m^*) \right] \tag{5.10}$$

In Equation 5.10, $\bar{V}_k \bar{V}_m^*$ can be written in the polar form as

$$\bar{V}_k \bar{V}_m^* = (V_k e^{j\theta_k})(V_m e^{-j\theta_m}) = V_k V_m e^{j\theta_{km}} = V_k V_m (\cos\theta_{km} + j\sin\theta_{km}) \tag{5.11}$$

where $\theta_{km} = \theta_k - \theta_m$.

Substituting Equation 5.11 into Equation 5.10 and separating real and imaginary parts (note that $\cos\theta_{kk} = 1$ and $\sin\theta_{kk} = 0$),

$$P_k = G_{kk}V_k^2 + V_k \sum_{\substack{m=1 \\ m \neq k}}^{n} V_m(G_{km}\cos\theta_{km} + B_{km}\sin\theta_{km}) \tag{5.12}$$

and

$$Q_k = -B_{kk}V_k^2 + V_k \sum_{\substack{m=1 \\ m \neq k}}^{n} V_m(G_{km}\sin\theta_{km} - B_{km}\cos\theta_{km}) \tag{5.13}$$

where the first terms on the right side of Equations 5.12 and 5.13 correspond to $m = k$. In an n-bus system, let us specify the types of buses as follows: one slack bus, n_{PV} PV buses, and n_{PQ} PQ buses such that

$$\underbrace{1}_{\text{slack-bus}} + n_{PV} + n_{PQ} = n \tag{5.14}$$

In this system, we have the following numbers of equations:

$(n_{PV} + n_{PQ})$ equations where P is specified
n_{PQ} equations where Q is specified

Therefore, there is a total of $(n_{PV} + 2n_{PQ})$ equations of a form similar to Equation 5.12 or 5.13.

Similarly, we have the following numbers of unknowns:

n_{PQ} unknown voltage magnitudes
$(n_{PV} + n_{PQ})$ unknown voltage phase angles

Therefore, there is a total of $(n_{PV} + 2n_{PQ})$ unknown variables to solve and the same number of equations.

There are no closed-form solutions to Equations 5.12 and 5.13 that are nonlinear; thus we will use a trial-and-error approach. Using Taylor's series expansion, we will linearize them and use an iterative procedure called the Newton-Raphson method until the solution is reached. To do this, we can write Equations 5.12 and 5.13 as follows, where P_k^{sp} and Q_k^{sp} are the specified values of injected real and reactive powers

$$P_k^{sp} - P_k(V_1,\ldots,V_n,\theta_1,\ldots,\theta_n) = 0 \ (k \equiv \text{all buses except the slack bus}) \tag{5.15}$$

and

$$Q_k^{sp} - Q_k(V_1,\ldots,V_n,\theta_1,\ldots,\theta_n) = 0 \ (k \equiv \text{all } PQ \text{ buses}) \tag{5.16}$$

5.6 NEWTON-RAPHSON PROCEDURE

To solve equations of the form given by Equations 5.15 and 5.16, the Newton-Raphson (N-R) procedure has become the commonly used approach due to its speed and the likelihood of conversion. Hence, we will focus on this procedure, recognizing that other procedures are also used, such as the Gauss-Seidel method described in Appendix A.

The N-R procedure is explained by means of a simple nonlinear equation of the same form as Equations 5.15 and 5.16

$$c - f(x) = 0 \qquad (5.17)$$

where c is a constant and $f(x)$ is a nonlinear function of a variable x. To determine the value of x that satisfies Equation 5.17, we start with an initial guess $x^{(0)}$ that is not exactly the solution but close to it. A small adjustment Δx is needed so that $(x^{(0)} + \Delta x)$ comes closer to the actual solution:

$$c - f(x^{(0)} + \Delta x) \approx 0 \qquad (5.18)$$

Using the Taylor series expansion, the function in Equation 5.18 can be expressed as follows, where the terms involving higher orders of Δx are ignored:

$$c - \left[f(x^{(0)}) + \frac{\partial f}{\partial x}\bigg|_0 \Delta x \right] = 0 \qquad (5.19)$$

or

$$c - f(x^{(0)}) = \frac{\partial f}{\partial x}\bigg|_0 \Delta x \qquad (5.20)$$

where the partial derivative is taken at $x = x^{(0)}$. The objective in Equation 5.20 is to calculate the adjustment Δx, which can be calculated as

$$\Delta x = \frac{c - f\left(x^{(0)}\right)}{\dfrac{\partial f}{\partial x}\bigg|_0} \qquad (5.21)$$

This results in the new estimate of x,

$$x^{(1)} = x^{(0)} + \Delta x \qquad (5.22)$$

Now, similar to Equation 5.21, the new correction to the estimate can be calculated as

$$\Delta x = \frac{c - f\left(x^{(1)}\right)}{\dfrac{\partial f}{\partial x}\bigg|_1} \qquad (5.23)$$

where the partial derivative is taken at $x = x^{(1)}$. Therefore,

$$x^{(2)} = x^{(1)} + \Delta x \qquad (5.24)$$

This process is repeated until the error $\varepsilon = |c - f(x)|$ is below a certain prespecified error tolerance value and the convergence is assumed to have been reached such that the original equation, Equation 5.17, is satisfied. This N-R procedure is illustrated in the following simple example.

Example 5.2
Consider a simple equation

$$4 - x^2 = 0 \qquad (5.25)$$

where, corresponding to Equation 5.17, $c = 4$ and $f(x) = x^2$. Solve for x that satisfies Equation 5.25 with an error ε below a tolerance of $0{:}0005$.

Solution The solution to this equation is obvious: $x = 2$. However, to illustrate the N-R procedure, the left side of Equation 5.25 is plotted in Figure 5.3 as a function of x.

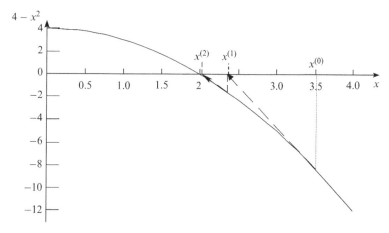

FIGURE 5.3 Plot of $4 - x^2$ as a function of x.

For $f(x) = x^2$, $\dfrac{\partial f}{\partial x} = 2x$. Using this partial derivative and the initial guess as $x^{(0)} = 3.5$, from Equation 5.21,

$$\Delta x = \frac{c - f\left(x^{(0)}\right)}{\left.\dfrac{\partial f}{\partial x}\right|_0} = -1.17857, \ x^{(1)} = x^{(0)} + \Delta x = 2.3214, \text{ and}$$

$$\varepsilon = \left|4 - (x^{(1)})^2\right| = 1.3890$$

Using $x^{(1)} = 2.3214$, from Equation 5.23,

$$\Delta x = \frac{c - f\left(x^{(1)}\right)}{\left.\dfrac{\partial f}{\partial x}\right|_1} = -0.29915, \quad x^{(2)} = x^{(1)} + \Delta x = 2.022, \text{ and}$$

$$\varepsilon = \left|4 - (x^{(2)})^2\right| = 0.0895$$

Repeating this procedure,

$$\Delta x = \frac{c - f\left(x^{(2)}\right)}{\left.\dfrac{\partial f}{\partial x}\right|_2} = -0.02188, \quad x^{(3)} = x^{(2)} + \Delta x = 2.000122,$$

$$\varepsilon = \left|4 - (x^{(3)})^2\right| = 0.00049$$

The error $\varepsilon = 0.00049$ is now below the prespecified tolerance, and the conversion is achieved.

5.7 SOLUTION OF POWER-FLOW EQUATIONS USING THE NEWTON-RAPHSON METHOD

Having looked at the basics of the N-R procedure, we are ready to apply it to power-flow Equations 5.15 and 5.16. In Equations 5.12 and 5.13, P_k and Q_k are in terms of the estimates of voltage magnitudes and phase angles, some of which are unknown and are thus yet to be determined. In a manner similar to Equation 5.20, we can write the following matrix equation, where corrections are expressed by ΔV and $\Delta\theta$:

$$\underbrace{\begin{bmatrix} P^{sp} - P \\ Q^{sp} - Q \end{bmatrix}}_{(2n_{PQ} + n_{PV}) \times 1} = \underbrace{\begin{bmatrix} \dfrac{\partial P}{\partial \theta} & \dfrac{\partial P}{\partial V} \\ \dfrac{\partial Q}{\partial \theta} & \dfrac{\partial Q}{\partial V} \end{bmatrix}}_{\substack{[J] \\ (2n_{PQ} + n_{PV}) \times (2n_{PQ} + n_{PV})}} \underbrace{\begin{bmatrix} \Delta\theta \\ \Delta V \end{bmatrix}}_{(2n_{PQ} + n_{PV}) \times 1} \quad (5.26)$$

The elements of the $[J]$ matrix, $\dfrac{\partial P}{\partial \theta}$ and so on, are submatrices, and $\Delta\theta$ and ΔV are vectors.

To evaluate the partial derivatives, we should recognize the following, noting that

$$\theta_{km} = \theta_k - \theta_m, \frac{\partial\left(\cos\theta_{km}\right)}{\partial\theta_k} = -\sin\theta_{km}, \frac{\partial\left(\cos\theta_{km}\right)}{\partial\theta_m} = \sin\theta_{km}, \frac{\partial\left(\sin\theta_{km}\right)}{\partial\theta_k}$$

$$= \cos\theta_{km} \text{ and } \frac{\partial\left(\sin\theta_{km}\right)}{\partial\theta_m} = -\cos\theta_{km}$$

The steps are as follows:

Step 1: Calculations of $\frac{\partial P}{\partial\theta}$ at all *PV* and *PQ* buses

At a bus k, from Equation 5.12, the partial derivatives with respect to (abbreviated wrt) θ for the real power are as follows:

$$\frac{\partial P_k}{\partial\theta_k} = V_k \sum_{\substack{m=1 \\ m\neq k}}^{n} V_m(-G_{km}\sin\theta_{km} + B_{km}\cos\theta_{km})\,(\text{wrt }\theta_k) \tag{5.27}$$

and

$$\frac{\partial P_k}{\partial\theta_j} = V_k V_j(G_{kj}\sin\theta_{kj} - B_{kj}\cos\theta_{kj})\,(\text{wrt }\theta_j, \text{ where } j\neq k) \tag{5.28}$$

Step 2: Calculations of $\frac{\partial P}{\partial V}$ at all *PQ* buses

At a bus k, from Equation 5.12, the partial derivatives with respect to V for the real power are as follows:

$$\frac{\partial P_k}{\partial V_k} = 2G_{kk}V_k + \sum_{\substack{m=1 \\ m\neq k}}^{n} V_m(G_{km}\cos\theta_{km} + B_{km}\sin\theta_{km})\,(\text{wrt }V_k) \tag{5.29}$$

and

$$\frac{\partial P_k}{\partial V_j} = V_k(G_{kj}\cos\theta_{kj} + B_{kj}\sin\theta_{kj})\,(\text{wrt }V_j, \text{ where } j\neq k) \tag{5.30}$$

Step 3: Calculations of $\frac{\partial Q}{\partial\theta}$ at all *PV* and *PQ* buses

At a bus k, from Equation 5.13, partial derivatives with respect to θ for the reactive power are as follows:

$$\frac{\partial Q_k}{\partial\theta_k} = V_k \sum_{\substack{m=1 \\ m\neq k}}^{n} V_m(G_{km}\cos\theta_{km} + B_{km}\sin\theta_{km})\,(\text{wrt }\theta_k) \tag{5.31}$$

and

$$\frac{\partial Q_k}{\partial \theta_j} = V_k V_j (-G_{kj} \cos\theta_{kj} - B_{kj} \sin\theta_{kj}) \, (\text{wrt}\,\theta_j, \text{ where } j \neq k) \qquad (5.32)$$

Step 4: Calculations of $\frac{\partial Q}{\partial V}$ at all PQ buses

At a bus k, from Equation 5.13, partial derivatives with respect to V for the reactive power are as follows:

$$\frac{\partial Q_k}{\partial V_k} = -2B_{kk}V_k + \sum_{\substack{m=1 \\ m \neq k}}^{n} V_m (G_{km} \sin\theta_{km} - B_{km} \cos\theta_{km}) \, (\text{wrt } V_k) \qquad (5.33)$$

and

$$\frac{\partial Q_k}{\partial V_j} = V_k (G_{kj} \sin\theta_{kj} - B_{kj} \cos\theta_{kj}) \, (\text{wrt } V_j, \text{ where } j \neq k) \qquad (5.34)$$

It should be noted that another row in Equation 5.26 is added for each PV bus that becomes a PQ bus due to its reactive power reaching one of its limits if they are specified. This is further explained in Section 5.10.

Convergence to the Correct Solution
The iterative N-R procedure is continued until all the mismatches in the left-side vector in Equation 5.26 are less than the specified tolerance values, at which point it is assumed that the solution has converged to the correct values. Until convergence is achieved, in each iterative step, the previous matrix equation is solved for the needed corrections, as are $\Delta\theta$s and ΔVs. In a practical power network with thousands of buses, the Jacobian J is calculated using sparsity and optimal ordering techniques. This discussion is beyond the scope of this book, and for the example system at hand, we will use the matrix inversion in Equation 5.26 to calculate the correction terms in each iteration step as follows:

$$\begin{bmatrix} \Delta\theta \\ \Delta V \end{bmatrix} = [J]^{-1} \begin{bmatrix} P^{sp} - P \\ Q^{sp} - Q \end{bmatrix} \qquad (5.35)$$

Example 5.3
In the example system in Figure 5.1, ignore all the shunt susceptances. Bus 1 is a slack bus, bus 2 is a PV bus, and bus 3 is a PQ bus. Using the N-R procedure described previously, assemble the Jacobian matrix for the example power system.

Solution   In this system, there are three buses with $n = 3$, $n_{PV} = 1$, and $n_{PQ} = 1$. There are three specified injected real/reactive powers (P_2^{sp}, P_3^{sp}, and Q_3^{sp}) and three unknowns (θ_2, θ_3, and V_3) related to the bus voltages. Therefore, the correction terms in Equation 5.35 are of the following form:

$$
\begin{bmatrix} P_2^{sp} - P_2 \\ P_3^{sp} - P_3 \\ Q_3^{sp} - Q_3 \end{bmatrix} = \underbrace{\begin{bmatrix} \dfrac{\partial P_2}{\partial \theta_2} & \dfrac{\partial P_2}{\partial \theta_3} & \dfrac{\partial P_2}{\partial V_3} \\[2mm] \dfrac{\partial P_3}{\partial \theta_2} & \dfrac{\partial P_3}{\partial \theta_3} & \dfrac{\partial P_3}{\partial V_3} \\[2mm] \dfrac{\partial Q_3}{\partial \theta_2} & \dfrac{\partial Q_3}{\partial \theta_3} & \dfrac{\partial Q_3}{\partial V_3} \end{bmatrix}}_{J} \begin{bmatrix} \Delta \theta_2 \\ \Delta \theta_3 \\ \Delta V_3 \end{bmatrix} \tag{5.36}
$$

To assemble the Jacobian matrix J of Equation 5.36, the bus values of $k, j,$ and m in the equations related to the N-R procedure can be recognized for each element, as shown in Table 5.2.

Equations for the Jacobian elements are as follows:

$$
\begin{aligned} J(1,1) &= V_2 V_1 (-G_{21} \sin \theta_{21} + B_{21} \cos \theta_{21}) \\ &\quad + V_2 V_3 (-G_{23} \sin \theta_{23} + B_{23} \cos \theta_{23}) \end{aligned} \tag{5.37}
$$

$$
J(1,2) = V_2 V_3 (G_{23} \sin \theta_{23} - B_{23} \cos \theta_{23}) \tag{5.38}
$$

$$
J(1,3) = V_2 (G_{23} \cos \theta_{23} + B_{23} \sin \theta_{23}) \tag{5.39}
$$

$$
J(2,1) = V_3 V_2 (G_{32} \sin \theta_{32} - B_{32} \cos \theta_{32}) \tag{5.40}
$$

$$
\begin{aligned} J(2,2) &= V_3 V_1 (-G_{31} \sin \theta_{31} + B_{31} \cos \theta_{31}) \\ &\quad + V_3 V_2 (-G_{32} \sin \theta_{32} + B_{32} \cos \theta_{32}) \end{aligned} \tag{5.41}
$$

$$
\begin{aligned} J(2,3) &= 2G_{33}V_3 + V_1 (G_{31} \cos \theta_{31} + B_{31} \sin \theta_{31}) \\ &\quad + V_2 (G_{32} \cos \theta_{32} + B_{32} \sin \theta_{32}) \end{aligned} \tag{5.42}
$$

$$
J(3,1) = V_3 V_2 (-G_{32} \cos \theta_{32} - B_{32} \sin \theta_{32}) \tag{5.43}
$$

$$
\begin{aligned} J(3,2) &= V_3 V_1 (G_{31} \cos \theta_{31} + B_{31} \sin \theta_{31}) \\ &\quad + V_3 V_2 (G_{32} \cos \theta_{32} + B_{32} \sin \theta_{32}) \end{aligned} \tag{5.44}
$$

$$
\begin{aligned} J(3,3) &= -2B_{33} V_3 + V_1 (G_{31} \sin \theta_{31} - B_{31} \cos \theta_{31}) \\ &\quad + V_2 (G_{32} \sin \theta_{32} - B_{32} \cos \theta_{32}) \end{aligned} \tag{5.45}
$$

Example 5.4

In the example system in Figure 5.1, ignore all the shunt susceptances. Bus 1 is a slack bus with $V_1 = 1.0$ pu and $\theta_1 = 0$. Bus 2 is a PV bus with $V_2 = 1.05$ pu and

TABLE 5.2 Buses related to the Jacobian matrix of Equation 5.36

Equation 5.27	Equation 5.28	Equation 5.30
$\dfrac{\partial P_2}{\partial \theta_2}: k = 2; m = 1, 3$	$\dfrac{\partial P_2}{\partial \theta_3}: k = 2, j = 3$	$\dfrac{\partial P_2}{\partial V_3}: k = 2; j = 3$
Equation 5.28	Equation 5.27	Equation 5.29
$\dfrac{\partial P_3}{\partial \theta_2}: k = 3; j = 2$	$\dfrac{\partial P_3}{\partial \theta_3}: k = 3; m = 1, 2$	$\dfrac{\partial P_3}{\partial V_3}: k = 3; m = 1, 2$
Equation 5.32	Equation 5.31	Equation 5.33
$\dfrac{\partial Q_3}{\partial \theta_2}: k = 3; j = 2$	$\dfrac{\partial Q_3}{\partial \theta_3}: k = 3; m = 1, 2$	$\dfrac{\partial Q_3}{\partial V_3}: k = 3; m = 1, 2$

$P_2^{sp} = 2.0$ pu. Bus 3 is a *PQ* bus with injections of $P_2^{sp} = -5.0$ pu and $Q_3^{sp} = -1.0$ pu. Using the previous N-R procedure, the Jacobian matrix was assembled in Example 5.3. Using that in the N-R procedure, calculate the power flow on all three lines in this example power system.

Solution   The Python program and MATLAB program files for this example are included on the accompanying website, and the results are as follows.

After the N-R procedure has converged, the Jacobian matrix is as follows:

$$J = \begin{bmatrix} 43.42 & -21.56 & 0.4237 \\ -21.07 & 35.99 & -1.5913 \\ 0.414 & -8.44 & 34.75 \end{bmatrix}$$

The bus voltages are as follows:

$$\bar{V}_1 = 1\angle 0° \text{ pu}, \quad \bar{V}_2 = 1.05\angle -2.07° \text{ pu}, \quad \bar{V}_3 = 0.978\angle -8.79° \text{ pu}$$

At bus 1, $P_{1-2} + jQ_{1-2} = (0.69 - j1.11)$ pu.

At bus 1, $P_{1-3} + jQ_{1-3} = (2.39 + j0.29)$ pu .

At bus 2, $P_{2-3} + jQ_{2-3} = (2.68 + j1.48)$ pu .

The power and reactive powers supplied by the generators at buses 1 and 2 are as follows:

$$P_1 + jQ_1 = (3.08 - j0.82) \text{pu} \quad \text{and} \quad P_2 + jQ_2 = (2.0 + j2.67) \text{pu}$$

These results are graphically shown in Figure 5.4, which can be verified by a commercial software package such as [4]. Graphical results from PSS®E are shown in Figure 5.5.

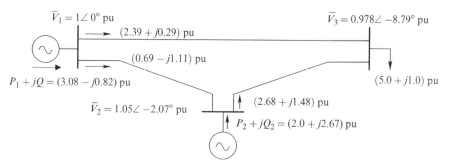

FIGURE 5.4   Power-flow results for Example 5.4.

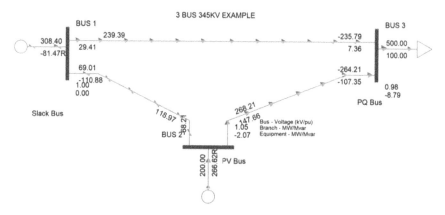

FIGURE 5.5   Power-flow results for Example 5.4 using PSS®E.

5.8   FAST DECOUPLED NEWTON-RAPHSON METHOD FOR POWER FLOW

In power systems, generally reactive powers Qs influence the voltage magnitudes, and the real powers Ps influence the phase angle θs. Therefore, the solution using the N-R method can be considerably simplified by including only the couplings mentioned and ignoring the $\partial P/\partial V$ and $\partial Q/\partial \theta$ terms in the Jacobian of Equation 5.26. Thus,

$$\left[P^{sp} - P\right] = \left[\frac{\partial P}{\partial \theta}\right]\left[\Delta \theta\right] \tag{5.46}$$

$$\left[Q^{sp} - Q\right] = \left[\frac{\partial Q}{\partial V}\right]\left[\Delta V\right] \tag{5.47}$$

These two equations are much faster to solve than the coupled set of equations in the full N-R procedure. Further simplifications can be made such that the elements of the Jacobian matrices in Equations 5.46 and 5.47 are constants and hence do not need to be calculated at every iteration, unlike the scenario in the full N-R method [5]. This technique can be very useful for calculating power flow for contingencies where the speed of calculations is of primary importance, even if the accuracy is somewhat sacrificed compared to the full N-R method.

5.9   SENSITIVITY ANALYSIS

The fast-decoupled formulation of Section 5.8 also shows that these equations can be used for sensitivity analysis: for example, to determine where to install a reactive power-supplying device to control the bus voltage magnitude. This can be seen by rewriting Equation 5.47 as

$$[\Delta Q] = \left[\frac{\partial Q}{\partial V}\right][\Delta V] \tag{5.48}$$

Thus,

$$[\Delta V] = \left[\frac{\partial Q}{\partial V}\right]^{-1}[\Delta Q] \tag{5.49}$$

Equation 5.49 demonstrates the sensitivity of various bus voltage magnitudes to the incremental change in reactive power at a selected bus.

5.10   REACHING THE BUS VAR LIMIT

As discussed in Chapter 9, synchronous generators have limits on the amount of reactive power (var) they can supply. Certain buses may contain additional devices to supply var, but such devices also have limits. Therefore, in a power-flow condition, if the demand of var at any PV bus reaches its limit, that bus voltage can be held at its specified magnitude and thus must be treated as a PQ bus.

For example, in the three-bus system shown in Figure 5.1, bus 2 is a PV bus. If the var needed from the generator at this bus reaches the limit, bus 2 must be treated as a PQ bus. Therefore, Equation 5.36 without the limit is modified as follows, where $Q_2^{\lim}$ is the var limit at bus 2, and a column and a row, shown in bold, must be added to the Jacobian matrix:

$$
\begin{bmatrix}
P_2^{sp} - P_2 \\
P_3^{sp} - P_3 \\
Q_3^{sp} - Q_3 \\
Q_2^{\lim} - Q_2
\end{bmatrix}
=
\underbrace{
\begin{bmatrix}
\dfrac{\partial P_2}{\partial \theta_2} & \dfrac{\partial P_2}{\partial \theta_3} & \dfrac{\partial P_2}{\partial V_3} & \dfrac{\partial P_2}{\partial V_2} \\[2mm]
\dfrac{\partial P_3}{\partial \theta_2} & \dfrac{\partial P_3}{\partial \theta_3} & \dfrac{\partial P_3}{\partial V_3} & \dfrac{\partial P_3}{\partial V_2} \\[2mm]
\dfrac{\partial Q_3}{\partial \theta_2} & \dfrac{\partial Q_3}{\partial \theta_3} & \dfrac{\partial Q_3}{\partial V_3} & \dfrac{\partial Q_3}{\partial V_2} \\[2mm]
\dfrac{\partial \boldsymbol{Q_2}}{\partial \boldsymbol{\theta_2}} & \dfrac{\partial \boldsymbol{Q_2}}{\partial \boldsymbol{\theta_3}} & \dfrac{\partial \boldsymbol{Q_2}}{\partial \boldsymbol{V_3}} & \dfrac{\partial \boldsymbol{Q_2}}{\partial \boldsymbol{V_2}}
\end{bmatrix}}_{J}
\begin{bmatrix}
\Delta \theta_2 \\
\Delta \theta_3 \\
\Delta V_3 \\
\Delta V_2
\end{bmatrix}
\tag{5.50}
$$

In an *n*-bus case, similar modifications are needed for all *PV* buses that reach their var limits and must be treated as *PQ* buses.

5.11 SYNCHRONIZED PHASOR MEASUREMENTS, PHASOR MEASUREMENT UNITS (PMUS), AND WIDE-AREA MEASUREMENT SYSTEMS

In digital relays, which are increasingly used to protect power systems as described in Chapter 13, it is possible to measure the phase angles of the bus voltages at the same instant. These synchronized phasor measurements, in phasor measurement units (PMUs) deployed over a large part of the power systems, known as *wide-area measurement systems*, can be used for control, monitoring, and protection. These are all part of the smart-grid initiative.

5.12 DC POWER FLOW

The decoupled N-R method is fast, but it is still an iterative algorithm and sometimes may face convergence issues. For many practical applications, computational speed is of high priority, and the accuracy of the results within an acceptable range is satisfactory. Some practical observations on the high-voltage transmission lines further simplify the power-flow equations [6] and also improve the computational speed by obtaining a linear one-step formulation. The power-flow equations are simplified based on the following observations:

1. In high-voltage transmission lines, the resistance is usually significantly smaller than the reactance. In other words, the ratio x/r is high.

    For any given transmission line, $y = g + jb, z = r + jx, y = \dfrac{1}{z}$.

    This can be written as $g + jb = \dfrac{r}{r^2 + x^2} - j\dfrac{x}{r^2 + x^2}$.

    Due to this assumption, $x >> r, g = 0, b = -\dfrac{1}{x}$. Since $g = 0$ for each transmission line, this means $G = 0$.

2. Under normal operating conditions, the difference in the voltage angles at two buses k and m connected by the transmission line are small.
    This means $\theta_k - \theta_m \approx 0$.
    It implies that $\cos(\theta_k - \theta_m) \approx 1, \sin(\theta_k - \theta_m) \approx \theta_k - \theta_m$.

3. In most cases, voltage magnitudes under normal operating conditions are within limits of 0.95 to 1.05 in a per-unit system. It is acceptable to assume the voltage magnitudes are very close to 1.0 per-unit. This implies $V_m = 1, V_k = 1$.

Revisiting the power-flow Equations 5.12 and 5.13 under these assumptions,

$$P_k = \sum_{\substack{m=1 \\ m \neq k}}^{n} B_{km}(\theta_k - \theta_m) \tag{5.51}$$

and

$$Q_k = 0 \tag{5.52}$$

Here, P and B are known variables and θ are unknown. Expanding these equations allows us to rewrite them in a matrix form as

$$P = -B\theta \tag{5.53}$$

Example 5.5
In the example system in Figure 5.4, using the DC power-flow procedure, calculate the power flow on all three lines in this example power system.

Solution   The Python program and MATLAB program files for this example are included on the accompanying website, and the results are as follows.
The bus voltage angles are as follows:

$$\theta_2 = -1.6°, \theta_3 = -8.7°$$

The power flowing through the transmission lines is as follows:
At bus 1, $P_{1-2} = 0.6\,\text{pu}$

At bus 1, $P_{1-3} = 2.4\,\text{pu}$

At bus 2, $P_{2-3} = 2.6\,\text{pu}$

REFERENCES

1. W. D. Stevenson. 1982. *Elements of Power System Analysis*, 4th ed. McGraw-Hill.
2. P. Kundur. 1994. *Power System Stability and Control*. McGraw-Hill.
3. G. Stagg and A. H. El-Abiad. 1968. *Computer Methods in Power System Analysis*. McGraw-Hill.
4. PSS®E high performance transmission planning and analysis software. https://new. siemens.com/global/en/products/energy/energy-automation-and-smart-grid/pss-software/pss-e.html.
5. B. Stott and D. Alsac. 1974. "Fast Decoupled Load Flow." *IEEE Transactions on Power Apparatus and Systems*. Vol. PAS-93, pp. 859–869.
6. A. J. Wood, B. F. Wollenberg, and G. B. Sheblé. 2013. *Power Generation, Operation, and Control*. John Wiley & Sons.

PROBLEMS

5.1   In Example 5.1, include the line susceptances, and construct the bus-admittance Y matrix.

5.2 Include the line susceptances in Example 5.4, and compare the results with the solution ignoring them.

5.3 In Example 5.4, ignore the $\partial P/\partial V$ and $\partial Q/\partial \theta$ terms in the Jacobian matrix for a decoupled N-R solution, and compare results and the iterations needed with the full N-R procedure.

5.4 In Example 5.4, calculate the sensitivity of reactive power to voltage at bus 3. Compare the result of using this sensitivity analysis to injecting 1.0 pu reactive power at bus 3 in the N-R solution to Example 5.4 and computing the increase in the bus 3 voltage.

5.5 In Example 5.4, demonstrate the effect of reducing reactive power demand at bus 3 to zero on the bus 3 voltage.

5.6 In Example 5.4, demonstrate the effect of series compensation in line 1–2 in the example three-bus system in Figure 5.1 on the line power flows and bus voltages, where the series reactance of the line is reduced by 50% by inserting a capacitor in series with line 1–2.

5.7 Compute the power flow in Example 5.4 using the Gauss-Seidel method described in the appendix.

PSS®E-BASED PROBLEMS

5.8 Calculate the power flow in Example 5.4.

5.9 In Problem 5.8, include the line susceptances, and compare the results with those of Problem 5.2.

5.10 In Problem 5.8, perform the decoupled N-R power flow, and compare the results with those of Problem 5.3.

5.11 Compare the results of series compensation with that of Problem 5.6.

5.12 In the system from Example 5.4, the capability to supply var at bus 2 is limited to 2 pu. Recalculate the power flow.

5.13 In Problem 5.8, perform the DC power flow, and compare the results with those of Problem 5.3.

APPENDIX 5A   GAUSS-SEIDEL PROCEDURE FOR POWER-FLOW CALCULATIONS

At a *PQ* bus k, from Equation 5.6,

$$\bar{I}_k = \sum_{m=1}^{n} Y_{km} \bar{V}_m \tag{A5.1}$$

From Equation 5.7,

$$\bar{I}_k = \frac{P_k - jQ_k}{\bar{V}_k^*} \tag{A5.2}$$

Substituting Equation A5.2 into Equation A5.1,

$$\frac{P_k - jQ_k}{\bar{V}_k^*} = \sum_{m=1}^{n} Y_{km} \bar{V}_m \qquad (A5.3)$$

Therefore, rearranging Equation A5.3,

$$\bar{V}_k = \frac{1}{Y_{kk}} \left[\frac{P_k - jQ_k}{\bar{V}_k^*} - \sum_{\substack{m=1 \\ m \neq k}}^{n} Y_{km} \bar{V}_m \right] \qquad (A5.4)$$

On the right-hand side of Equation A5.4, P_k and Q_k are specified, and the voltage values are the original estimates to calculate the new value $\bar{V}_k$. New estimated values are used as soon as they are available.

At a PV bus k, where P_k and the voltage magnitude V_k are specified, from Equation A5.3,

$$P_k - jQ_k = \bar{V}_k^* \sum_{m=1}^{n} Y_{km} \bar{V}_m \qquad (A5.5)$$

Therefore, using the latest voltage estimates,

$$Q_k = -Im \left[\bar{V}_k^* \sum_{m=1}^{n} Y_{km} \bar{V}_m \right] \qquad (A5.6)$$

The value of Q_k calculated from this equation is used in Equation A5.4 to get the new estimate of the phase angle of $\bar{V}_k$, which, keeping the specified voltage magnitude V_k at this PV bus, gives us the new estimate of $\bar{V}_k$. If the value of Q_k calculated by Equation A5.6 is outside the range of the minimum and the maximum reactive power that can be supplied at this bus, the limit being violated is used in Equation A5.4.

This procedure is repeated until the solution converges.

APPENDIX 5B   REMOTE BUS VOLTAGE CONTROL BY GENERATORS

To illustrate this concept, consider an extended version of the three-bus system in Figure 5.1. Specifically, the generator on bus 2 is moved to the new bus 4 and regulates bus 2. The transmission line from bus 2 to bus 4 has an impedance of $0.00047 + j0.00474$. This is illustrated in Figure 5.6.

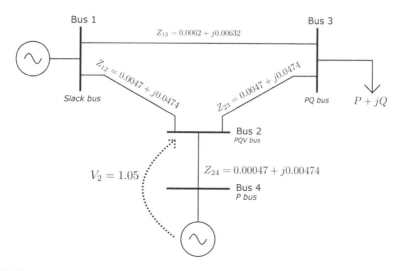

FIGURE 5.6 A four-bus 345 kV example system.

Since the generator at bus 4 regulates the voltage of bus 2, it introduces some new notions of bus types:

1. *P*-type bus [bus 4 in Figure 5.6]: a generator bus where the voltage magnitude *V* and the power *P* are specified, but instead of regulating the voltage of its own bus, it regulates the voltage of a remote bus.
2. *PQV*-type bus [bus 2 in Figure 5.6]: a *PQ* bus where the voltage is specified via the remote control of a generator. In some literature, these are referred to as *PVQ* buses, but for this book, we use *PQV*.

With these new definitions, the known and unknown quantities can be summarized as follows:

Bus type	P	Q	V	θ
Slack (bus 1)	Unknown	Unknown	1	0
PQ (bus 3)	Known	Known	Unknown	Unknown
PQV (bus 2)	0	0	Known	Unknown
P (bus 4)	Known	Unknown	Unknown	Unknown

In this system, there are four buses with $n = 4$, $n_P = 1$, $n_{PQ} = 1$, and $n_{PQV} = 1$. There are three specified injected real/reactive powers (P_4^{sp}, P_3^{sp}, and Q_3^{sp}) and five unknowns (θ_2, θ_3, θ_4, V_3, and V_4) related to the bus voltages. Therefore, the correction terms in Equation 5.35 are of the following form:

$$\begin{bmatrix} P_2^{sp} - P_2 \\ P_3^{sp} - P_3 \\ P_4^{sp} - P_4 \\ Q_2^{sp} - Q_2 \\ Q_3^{sp} - Q_3 \end{bmatrix} = \underbrace{\begin{bmatrix} \dfrac{\partial P_2}{\partial \theta_2} & \dfrac{\partial P_2}{\partial \theta_3} & \dfrac{\partial P_2}{\partial \theta_4} & \dfrac{\partial P_2}{\partial V_3} & \dfrac{\partial P_2}{\partial V_4} \\ \dfrac{\partial P_3}{\partial \theta_2} & \dfrac{\partial P_3}{\partial \theta_3} & \dfrac{\partial P_3}{\partial \theta_4} & \dfrac{\partial P_3}{\partial V_3} & \dfrac{\partial P_3}{\partial V_4} \\ \dfrac{\partial P_4}{\partial \theta_2} & \dfrac{\partial P_4}{\partial \theta_3} & \dfrac{\partial P_4}{\partial \theta_4} & \dfrac{\partial P_4}{\partial V_3} & \dfrac{\partial P_4}{\partial V_4} \\ \dfrac{\partial Q_2}{\partial \theta_2} & \dfrac{\partial Q_2}{\partial \theta_3} & \dfrac{\partial Q_2}{\partial \theta_4} & \dfrac{\partial Q_2}{\partial V_3} & \dfrac{\partial Q_2}{\partial V_4} \\ \dfrac{\partial Q_3}{\partial \theta_2} & \dfrac{\partial Q_3}{\partial \theta_3} & \dfrac{\partial Q_3}{\partial \theta_4} & \dfrac{\partial Q_3}{\partial V_3} & \dfrac{\partial Q_3}{\partial V_4} \end{bmatrix}}_{J} \begin{bmatrix} \Delta \theta_2 \\ \Delta \theta_3 \\ \Delta \theta_4 \\ \Delta V_3 \\ \Delta V_4 \end{bmatrix} \qquad (B5.1)$$

Some sample equations for the Jacobian elements are as follows:

$$\begin{aligned} J(1,1) &= V_2 V_1 \left(-G_{21} \sin \theta_{21} + B_{21} \cos \theta_{21} \right) \\ &+ V_2 V_3 \left(-G_{23} \sin \theta_{23} + B_{23} \cos \theta_{23} \right) \\ &+ V_2 V_4 \left(-G_{24} \sin \theta_{24} + B_{24} \cos \theta_{24} \right) \end{aligned} \qquad (B5.2)$$

$$J(1,2) = V_2 V_3 (G_{23} \sin \theta_{23} - B_{23} \cos \theta_{23}) \qquad (B5.3)$$

$$J(1,3) = V_2 V_4 (G_{24} \sin \theta_{24} - B_{24} \cos \theta_{24}) \qquad (B5.4)$$

$$J(1,4) = V_2 (G_{23} \cos \theta_{23} + B_{23} \sin \theta_{23}) \qquad (B5.5)$$

$$J(1,5) = V_2 (G_{24} \cos \theta_{24} + B_{24} \sin \theta_{24}) \qquad (B5.6)$$

Example 5.6
In the example system in Figure 5.6, ignore all the shunt susceptances. Bus 1 is a slack bus with $V_1 = 1.0$ pu and $\theta_1 = 0$. Bus 2 is a PQV bus with $V_2 = 1.05$ pu. Bus 4 is the P bus with $P_4^{sp} = 2.0$ pu. Bus 3 is a PQ bus with injections of $P_2^{sp} = -5.0$ pu and $Q_3^{sp} = -1.0$ pu. Using the N-R procedure, calculate the power flow on all four lines in this example power system.

Solution   The Python program and MATLAB program files for this example are included on the accompanying website, and the results are as follows.
    After the N-R procedure has converged, the Jacobian matrix is as follows:

$$J = \begin{bmatrix} 276.52 & -21.56 & -233.1 & 0.422 & -23.21 \\ -21.07 & 35.99 & 0 & -1.59 & 0 \\ -233.44 & 0 & 233.44 & 0 & 23.74 \\ -27.21 & -4.6 & -21.23 & -22.04 & -219.29 \\ 0.41 & -8.44 & 0 & 34.75 & 0 \end{bmatrix}$$

The bus voltages are as follows:

$$\overline{V}_1 = 1\angle 0° \text{ pu, } \overline{V}_2 = 1.05\angle -2.07° \text{ pu, } \overline{V}_3$$
$$= 0.978\angle -8.79° \text{ pu, } \overline{V}_4 = 1.06\angle -1.65° \text{ pu}$$

At bus 1, $P_{1\text{-}2} + jQ_{1\text{-}2} = (0.69 - j1.11)\text{pu}$

At bus 1, $P_{1\text{-}3} + jQ_{1\text{-}3} = (2.39 + j0.29)\text{pu}$

At bus 2, $P_{2\text{-}3} + jQ_{2\text{-}3} = (2.68 + j1.48)\text{pu}$

At bus 4, $P_{4\text{-}2} + jQ_{4\text{-}2} = (2.00 + j2.71)\text{pu}$

The power and reactive powers supplied by the generators at buses 1 and 4 are as follows:

$$P_1 + jQ_1 = (3.08 - j0.82)\text{pu and } P_4 + jQ_4 = (2.0 + j2.71)\text{pu}$$

Graphical results from PSS®E are shown in Figure 5.7.

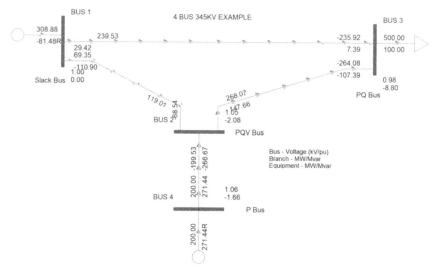

FIGURE 5.7 Power-flow results for Example 5.5 using PSS®E.

6

TRANSFORMERS IN POWER SYSTEMS

6.1 INTRODUCTION

Transformers are absolutely essential in making large-scale power transfer feasible over long distances. The primary role of transformers is to change the voltage level. For example, generation in power systems, primarily by synchronous generators, takes place at around the 20 kV level. However, this voltage is too low to transmit an economically significant amount of power over long distances. Therefore, transmission voltages of 230 kV, 345 kV, and 500 kV are common, and some are as high as 765 kV. At the load end, these voltages are stepped down to manageable and safe levels, such as 120/240 V single-phase in residential usage. Another reason for using transformers in many applications is to provide electrical isolation for safety purposes. Transformers are also needed for converters used in high-voltage DC transmission systems, as discussed in Chapter 7.

Typically, in power systems, voltages are transformed approximately five times between generation and delivery to the ultimate users. Hence, the total installed MVA ratings of transformers are as much as five times larger than those of generators.

6.2 BASIC PRINCIPLES OF TRANSFORMER OPERATION

Transformers consist of two or more tightly coupled windings where almost all the flux produced by one winding links the other windings. To understand the

Electric Power Systems with Renewables: Simulations Using PSS®E, Second Edition. Ned Mohan and Swaroop Guggilam.
© 2023 John Wiley & Sons, Inc. Published 2023 by John Wiley & Sons, Inc.
Companion Website: www.wiley.com/go/mohaneps

operating principles of transformers, consider a single coil, also called a *winding*, of N_1 turns, as shown in Figure 6.1a.

(a)   (b)

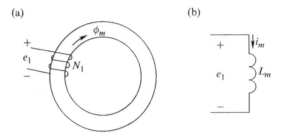

FIGURE 6.1   Principle of transformers, beginning with just one coil. (a) Core with winding N_1; (b) circuit representation.

Initially, we will assume that the resistance and leakage inductance of this winding are both zero; assuming zero leakage flux implies that all the flux produced by this winding is confined to the core. Applying a time-varying voltage e_1 to this winding results in a flux $\phi_m(t)$. From Faraday's law,

$$e_1(t) = N_1 \frac{d\phi_m}{dt} \tag{6.1}$$

where $\phi_m(t)$ is completely dictated by the time-integral of the applied voltage, as given here (where it is assumed that the flux in the winding is initially zero):

$$\phi_m(t) = \frac{1}{N_1} \int_0^t e_1(\tau) \cdot d\tau \tag{6.2}$$

6.2.1   Transformer Exciting Current

Transformers use ferromagnetic materials that guide magnetic flux lines and, due to their high permeability, require small ampere-turns (a small current for a given number of turns) to produce the desired flux density. These materials exhibit the multivalued nonlinear behavior shown by their *B-H* characteristics in Figure 6.2a.

Imagine that the toroid in Figure 6.1a consists of a ferromagnetic material such as silicon steel. If the current through the coil is slowly varied in a sinusoidal manner with time, the corresponding *H*-field will cause one of the hysteresis loops shown in Figure 6.2a to be traced. Completing the loop once results in a net dissipation of energy within the material, causing power loss referred to as *hysteresis loss*. Increasing the peak value of the sinusoidally varying *H*-field will

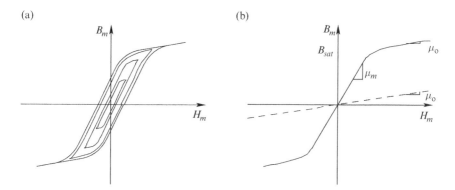

FIGURE 6.2 (a) *B-H* characteristics of ferromagnetic materials; (b) approximation of *B-H* characteristics.

result in a bigger hysteresis loop. Joining the peaks of the hysteresis loops, we can approximate the *B-H* characteristic by the single curve shown in Figure 6.2b. In Figure 6.2b, the linear relationship (with a constant μ_m) is approximately valid until the "knee" of the curve is reached, beyond which the material begins to saturate. Ferromagnetic materials are often operated up to a maximum flux density, slightly above the knee of 1.6 T to 1.8 T, beyond which many more ampere-turns are required but the flux-density increases only slightly. In the saturated region, the incremental permeability of the magnetic material approaches μ_o, as shown by the slope of the curve in Figure 6.2b.

In the *B-H* curve of the magnetic material in Figure 6.2b, in accordance with Faraday's law, B_m is proportional to the flux-linkage λ_m ($= N_1\phi_m$) of the coil; and in accordance with Ampere's law, H_m is proportional to the magnetizing current i_m drawn by the coil to establish this flux. Therefore, a plot similar to the *B-H* plot in Figure 6.2b can be drawn in terms of λ_m and i_m. In the linear region with a constant slope μ_m, this linear relationship between λ_m and i_m can be expressed by a magnetizing inductance L_m

$$L_m = \frac{\lambda_m}{i_m} \tag{6.3}$$

The current $i_m(t)$ drawn to establish this flux depends on the magnetizing inductance L_m of the winding, as depicted in Figure 6.1b. With a sinusoidal voltage applied to the winding as $e_1 = \sqrt{2}E_1(\text{rms})\cos\omega t$, the core flux from Equation 6.2 is $\phi_m = \hat{\phi}_m \sin\omega t$. In the phasor domain, the relationship between these two, in accordance with Faraday's law, can be expressed as

$$\sqrt{2}E_1(\text{rms}) = (2\pi f)N_1\hat{\phi}_m \tag{6.4}$$

or

$$\hat{\phi}_m \simeq \frac{E_1(\mathrm{rms})}{4.44 N_1 f} \tag{6.5}$$

which clearly shows that exceeding the applied voltage E_1 (rms) above its rated value will cause the maximum flux $\hat{\phi}_m$ to enter the saturation region in Figure 6.2b, resulting in excessive magnetizing current to be drawn. In the saturation region, the magnetizing current drawn is also distorted, with a significant amount of the third-harmonic component.

In modern power transformers with large kVA ratings, the magnetizing currents in their normal operating region are very small – for example, well below 0.2% of the rated current at the rated voltage.

6.2.2 Voltage Transformation

A second winding of N_2 turns is now placed on the core, as shown in Figure 6.3a. A voltage is induced in the second winding due to the time-varying flux $\phi_m(t)$ linking it.

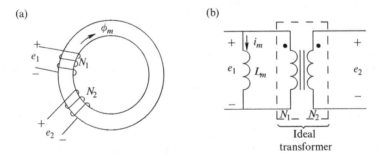

FIGURE 6.3 Transformer with an open-circuited second coil. (a) Core with windings N_1 and N_2; (b) circuit representation.

From Faraday's law,

$$e_2(t) = N_2 \frac{d\phi_m}{dt} \tag{6.6}$$

Equations 6.1 and 6.6 show that in each winding, the volts per turn are the same due to the same $d\phi_m/dt$:

$$\frac{e_1(t)}{N_1} = \frac{e_2(t)}{N_2} \tag{6.7}$$

We can represent the relationship of Equation 6.7 in Figure 6.3b by means of a hypothetical circuit component called the *ideal transformer*, which relates the voltages in the two windings by the turns ratio N_1/N_2 on an instantaneous basis or in terms of phasors in a sinusoidal steady state (shown in brackets):

$$\frac{e_1(t)}{e_2(t)} = \frac{N_1}{N_2} \quad \text{and} \quad \left(\frac{\bar{E}_1}{\bar{E}_2} = \frac{N_1}{N_2}\right) \tag{6.8}$$

The dots in Figure 6.3b convey the information that the winding voltages will have the same polarity at the dotted terminals with respect to their undotted terminals. For example, if ϕ_m is increasing with time, the voltages at both dotted terminals will be positive with respect to the corresponding undotted terminals. The advantage of using this dot convention is that the winding orientations on the core need not be shown in detail.

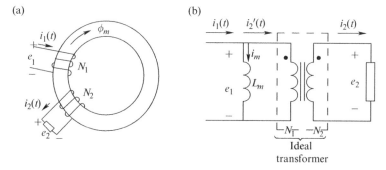

FIGURE 6.4 (a) Transformer with a load connected to the secondary winding; (b) circuit representation.

A load such as an *R-L* combination is now connected across the secondary winding, as shown in Figure 6.4a. A current $i_2(t)$ flows through the load. The resulting ampere-turns $N_2 i_2$ will tend to change the core flux ϕ_m but *cannot* because $\phi_m(t)$ is completely dictated by the time integral of the applied voltage $e_1(t)$, as given in Equation 6.2. Therefore, additional current i_2' in Figure 6.4b is drawn by winding 1 to compensate for (or nullify) $N_2 i_2$, such that $N_1 i_2' = N_2 i_2$:

$$\frac{i_2'(t)}{i_2(t)} = \frac{N_2}{N_1} \quad \text{and} \quad \left(\frac{\bar{I}_2'}{\bar{I}_2} = \frac{N_2}{N_1}\right) \tag{6.9}$$

This is the second property of the ideal transformer. Thus, the total current drawn from the terminals of winding 1 is

$$i_1(t) = i_m(t) + i_2'(t) \quad \text{and} \quad (\bar{I}_1 = \bar{I}_m + \bar{I}_2') \tag{6.10}$$

6.2.3　Transformer Equivalent Circuit

In Figure 6.4b, the resistance and leakage inductance associated with winding 2 appear in series with the *R-L* load. Therefore, the induced voltage e_2 differs from the voltage v_2 at the winding terminals by the voltage drop across the winding resistance and the leakage reactance, as depicted in Figure 6.5 in the phasor domain. Similarly, the applied voltage v_1 differs from the emf e_1 (induced by the time-rate of change of the flux ϕ_m) in Figure 6.4b by the voltage drop across the resistance and the leakage inductance of winding 1, represented in Figure 6.5 in the phasor domain.

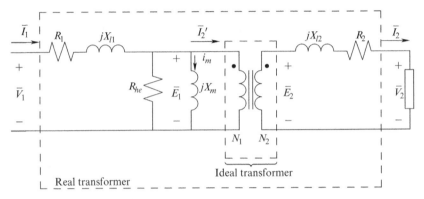

FIGURE 6.5　Transformer equivalent circuit including leakage impedances and core losses.

6.2.4　Core Losses

The loss due to the hysteresis loop in the *B-H* characteristic of the magnetic material was discussed earlier. Another source of core loss is due to eddy currents. All magnetic materials have a finite electrical resistivity (ideally, it should be infinite). By Faraday's voltage induction law, time-varying fluxes induce voltages in the core, which result in circulating (eddy) currents within the core to oppose these flux changes (and partially neutralize them).

In Figure 6.6a, an increasing flux ϕ will set up many current loops (due to induced voltages that oppose the change in core flux), which result in losses. The primary means of limiting the eddy-current losses is to make the core out of steel laminations insulated from each other by thin layers of varnish, as shown in Figure 6.6b.

A few laminations are shown to illustrate how insulated laminations reduce eddy-current losses. Because of the insulation between laminations, the current is forced to flow in much smaller loops within each lamination. Laminating the core reduces the flux and the induced voltage more than it

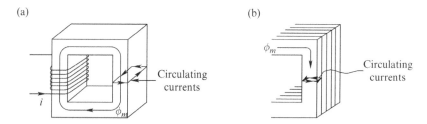

FIGURE 6.6 Eddy currents in the transformer core. (a) Without laminations; (b) with laminations.

reduces the effective resistance to the currents within a lamination, thus reducing the overall losses. For 50 or 60 Hz operation, lamination thicknesses are typically 0.2–1 mm. We can model core losses due to hysteresis and eddy currents by connecting a resistance R_{he} in parallel with X_m, as shown in Figure 6.5. In large modern transformers, core losses are well below 0.1% of the transformer MVA rating.

6.2.5 Equivalent Circuit Parameters

To utilize the transformer equivalent circuit in Figure 6.5, we need the values of its various parameters. These specifications are generally provided by the manufacturers of power transformers. They can also be obtained using short-circuit and open-circuit tests. In the open-circuit test, the rated voltage is applied to the low-voltage winding, keeping the high side open-circuited. This allows the magnetizing reactance and the core-loss equivalent resistance to be estimated. In the short-circuit test, the low-voltage winding is short-circuited, and a reduced voltage is applied to the high-voltage winding that results in the rated current. This allows the leakage impedances in the transformer equivalent circuit to be estimated. A detailed discussion of the open-circuit and short-circuit tests can be found in any basic book dealing with transformers.

6.3 SIMPLIFIED TRANSFORMER MODEL

Consider the equivalent circuit of a real transformer, shown in Figure 6.5. In many power system studies, the excitation current, which is the sum of the magnetizing current and the core-loss current components, is neglected, resulting in the simplified model shown in Figure 6.7. In this model, the subscript p refers to the primary winding and s to the secondary winding. Z_p and Z_s are the leakage impedances of the primary and the secondary windings, and the turns ratio $n = n_s/n_p$.

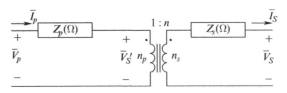

FIGURE 6.7   Simplified transformer model.

6.3.1   Transferring Leakage Impedances across the Ideal Transformer Portion

In the simplified model in Figure 6.7, if the secondary-winding terminals are hypothetically short-circuited, then $\bar{V}_s = 0$; and related by the ideal transformer turns ratio,

$$\bar{V}'_s = \frac{Z_s \bar{I}_s}{n} \quad \text{(under short-circuit)} \tag{6.11}$$

Also related by the ideal transformer turns ratio, $\bar{I}_s = \bar{I}_p/n$. Substituting for $\bar{I}_s$ in Equation 6.11,

$$\bar{V}'_s = \left(\frac{Z_s}{n^2}\right)\bar{I}_p \quad \text{(under short-circuit)} \tag{6.12}$$

In Figure 6.7,

$$\bar{V}_p = \bar{V}'_s + Z_p \bar{I}_p \tag{6.13}$$

and therefore, from the primary-winding terminals under this hypothetical short-circuit, the impedance "seen" from the primary side is $\bar{V}_p/\bar{I}_p$, which by using Equations 6.12 and 6.13 is

$$Z_{ps}(\Omega) = Z_p + (Z_s/n^2) \tag{6.14a}$$

as shown in Figure 6.8a. Similarly, if the primary-winding leakage impedance is transferred to the secondary winding, then

$$Z_{sp}(\Omega) = Z_s + (n^2 Z_p) \tag{6.14b}$$

as shown in Figure 6.8b.

(a)

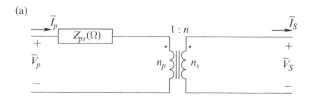

(b)

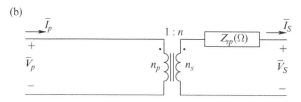

FIGURE 6.8 Transferring leakage impedances across the ideal portion of the transformer model. (a) Leakage impedance on the primary side of the transformer; (b) leakage impedance on the secondary side of the transformer.

6.4 PER-UNIT REPRESENTATION

Most power system studies, such as power flow, transient analysis, and short-circuit fault calculations, are carried out in terms of per-unit. In a transformer, the rated voltages and currents on each side are considered the base values. Since the voltages and currents on the two sides are related by the turns ratio, the MVA base, which is the product of the voltage and current bases, is the same on each side. In terms of base values, the base impedance magnitudes $Z_{p,\text{base}}$ and $Z_{s,\text{base}}$ are

$$Z_{p,\text{base}} = V_{p,\text{rated}} / I_{p,\text{rated}} \quad \text{and} \quad Z_{s,\text{base}} = V_{s,\text{rated}} / I_{s,\text{rated}} \tag{6.15}$$

The magnitudes of the voltage and currents associated with the ideal transformer in Figure 6.7 are related as follows:

$$\frac{V_{p,\text{rated}}}{V_{s,\text{rated}}} = \frac{1}{n} \quad \text{and} \quad \frac{I_{p,\text{rated}}}{I_{s,\text{rated}}} = n \tag{6.16}$$

Therefore, the base impedance magnitudes on the two sides, given by Equation 6.15, are related as

$$\frac{Z_{p,\text{base}}}{Z_{s,\text{base}}} = \left(\frac{1}{n}\right)^2 = \left(\frac{n_p}{n_s}\right)^2 \tag{6.17}$$

Using these base values, all the parameters and variables in Figures 6.8a and b can be expressed in per-unit as the ratio of their base values, as shown by a common equivalent circuit in Figure 6.9, where Z_{tr} (pu) is the transformer per-unit leakage impedance equal to Z_{ps} (pu) and Z_{sp} (pu):

$$Z_{tr}(\text{pu}) = \underbrace{Z_{ps}(\text{pu})}_{\left(=\dfrac{Z_{ps}}{Z_{p,base}}\right)} = \underbrace{Z_{sp}(\text{pu})}_{\left(=\dfrac{Z_{sp}}{Z_{s,base}}\right)} \qquad (6.18)$$

In Figure 6.9, the primary and secondary winding currents are equal in per-unit – that is, $\bar{I}_p(\text{pu}) = \bar{I}_s(\text{pu}) = \bar{I}(\text{pu})$ – and the voltages on the two sides differ by the voltage drop across the leakage impedance.

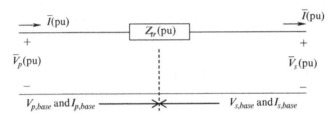

FIGURE 6.9   Transformer equivalent circuit in per-unit (pu).

In three-phase power systems, transformer windings are connected in a Y or delta arrangement, as shown in Figure 6.10. Example 6.1 shows their per-phase representation in per-unit, assuming that the primary-side and secondary-side windings are connected as Y-Y or delta-delta. If one side is Y-connected and the other is delta-connected, there is a phase shift of 30 degrees that must be taken into account, as discussed later.

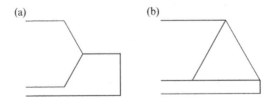

FIGURE 6.10   Winding connections in a three-phase system. (a) Y-connected; (b) delta-connected.

Example 6.1

Suppose the 200 km transmission line between buses 1 and 3 in the three-bus power system in Figure 5.1 in Chapter 5 is at 500 kV. Two 345/500 kV transformers are used at both ends, as shown in the one-line diagram in Figure 6.11. In the per-unit study, if the system base voltage is 345 kV, then the line impedance and transformer leakage impedances are calculated on the 345 kV voltage base and 100 MVA base. For this 500 kV transmission line, the series reactance is 0.326 Ω/km at 60 Hz, and the series resistance is 0.029 Ω/km. Neglect line susceptances. Each transformer has a leakage reactance of 0.2 pu on the 1000 MVA base.

(a)

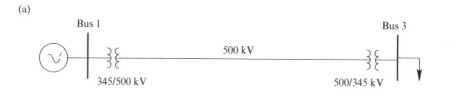

(b)

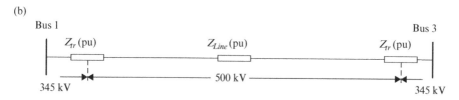

FIGURE 6.11 Including nominal-voltage transformers in per-unit. (a) Single-line diagram for Example 6.1; (b) per-unit representation.

Calculate the per-unit series impedance of the transmission line and the transformers between buses 1 and 3 for the power-flow study as discussed in Chapter 5, where the line-line voltage base is 345 kV and the three-phase MVA base is 100 MVA.

Solution The 500 kV transmission line is 200 km long. From the given parameter values, the series impedance of the line is $Z_{\text{Line}} = (5.8 + j65.2)$ Ω. In a three-phase system,

$$Z_{\text{base}}(\Omega) = \frac{kV_{\text{base}}^2 \, (\text{L-L})}{\text{MVA}_{\text{base}} \, (\text{3-phase})} \tag{6.19}$$

and therefore, at the 500 kV voltage and 1,000 MVA (three-phase) basis, the base impedance is $Z_{\text{base}} = 250.0$ Ω. Therefore, in per-unit, the series impedance of the transmission line is $Z_{\text{Line}} = (0.0232 + j0.2608)$ pu. Each of the transformer impedances is given as $Z_{\text{tr}} = j0.2$ pu. All the impedances in per-unit, on a 1000 MVA basis, are shown in Figure 6.11b.

Using Equation 6.18, the transmission-line impedance can be represented on the 345 kV side of either transformer, and its per-unit value will not change. Therefore, the impedance between buses 1 and 3 on a 345 kV, 1000 MVA base is as follows in the diagram in Figure 6.11b:

$$Z_{13} = j0.2 + (0.0232 + j0.2608) + j0.2 = (0.0232 + j0.6608) \text{ pu}$$

We now need to express this impedance on a 345 kV, 100 MVA base for use in the power-flow studies from Chapter 5. Using Equation 6.19, the per-unit impedance from an original MVA base to a new MVA base is as follows:

$$Z_{pu}(\text{new}) = Z_{pu}(\text{original}) \times \frac{\text{MVA}_{base}(\text{new})}{\text{MVA}_{base}(\text{original})} \tag{6.20}$$

Therefore, from Equation 6.20, using 100 MVA as the new base and 1000 MVA as the original base, the series impedance between buses 1 and 3 is

$$Z_{13} = (0.00232 + j0.06608) \text{ pu}$$

The power-flow results using transformer representation and the equivalent imped-ance are performed using PSS®E, and the results are shown in Figure 6.12 and Figure 6.13. The PSS®E files for this example are included on the accompanying website. As can be seen, the power-flow results match closely.

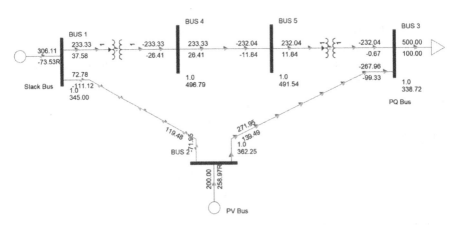

FIGURE 6.12 Power flow of the example system with two transformers.

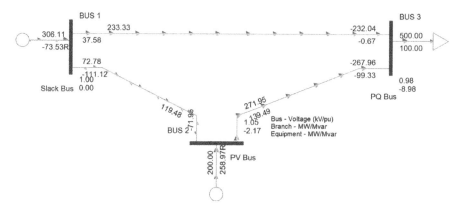

FIGURE 6.13 Power flow of the example system with two transformers represented as an equivalent series impedance.

6.5 TRANSFORMER EFFICIENCIES AND LEAKAGE REACTANCES

Power transformers are designed to minimize power losses within them. These consist of the I^2R losses, also called the *copper losses*, in the windings, and the core losses. Core losses are mostly independent of the transformer loading, whereas the losses in windings depend on the square of the transformer loading. The energy efficiency of a transformer or any piece of equipment is defined as follows, where the output power equals the input power minus the losses:

$$\%\text{efficiency} = 100 \times \frac{P_{\text{output}}}{P_{\text{input}}} = 100 \times \left(1 - \frac{P_{\text{losses}}}{P_{\text{input}}}\right) \qquad (6.21)$$

Generally, transformer efficiencies are at their maximum at a load when the core and winding losses are equal to each other. In large power transformers, these efficiencies generally exceed 99.5% at or near full load.

To achieve high efficiencies in transformers, their winding resistances are generally well below 0.5% or 0.005 pu. Therefore, their leakage impedances are dominated by the leakage reactances, which depend on the voltage class of the transformer. These leakage reactances are approximately in the following ranges: 7–10% in 69 kV transformers, 8–12% in 115 kV transformers, and 11–16% in 230 kV transformers. In 345 kV and 500 kV transformers, these reactances are 20% or even larger.

6.6 REGULATION IN TRANSFORMERS

Large values of leakage reactances help reduce currents during power-system faults like short-circuits on transmission lines to which these transformers are connected. However, for a constant input voltage applied, the output voltage of a transformer changes based on the transformer loading, due to the voltage drop across its leakage impedance. This is called *regulation*, which is the change in output voltage expressed as the percentage of the rated output voltage if the rated output kVA load at a specified power factor is reduced to zero (that is, the secondary winding is open-circuited). From Figure 6.9, assuming the leakage impedance to be purely reactive,

$$\bar{V}_s(\text{pu}) = \bar{V}_P(\text{pu}) - jX_{tr}(\text{pu})\bar{I}(\text{pu}) \tag{6.22}$$

It can be observed from Equation 6.22 that the lower the power factor (lagging) of the load, the larger the change in the output voltage magnitude and hence the larger the regulation.

6.6.1 Transformer Tap-Changing for Voltage Control

Tap-changing in transformers makes it possible to adjust the output voltage magnitude. It is possible to change taps under load, and such arrangements, called load tap changers (LTCs), are described in detail in [1]. Tap-changing is usually accomplished using autotransformers, discussed in the next section. Tap-changing can be included in power-flow studies, as illustrated in Problem 6.17 using PSS®E.

6.7 AUTOTRANSFORMERS

Autotransformers are very frequently used in power systems for transforming voltages where electrical isolation is not necessary and for tap-changing. In this analysis, we will assume an ideal transformer, neglecting the leakage impedances and excitation current. Therefore, all the voltages and currents are of the same phase, respectively, and hence are represented by their magnitudes alone for convenience rather than as phasors.

In the two-winding arrangement shown in Figure 6.14a, in terms of the rated voltage and the rated current associated with each winding, the transformer rating is

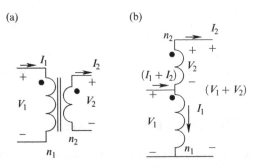

FIGURE 6.14 (a) Isolated winding transformer; (b) autotransformer.

$$\text{two-winding transformer rating} = V_1 I_1 = V_2 I_2 \tag{6.23}$$

In the autotransformer arrangement shown in Figure 6.14b, these two winding are connected in series. With V_1 applied to the low side on the left, the high-side voltage is $(V_1 + V_2)$, as shown in Figure 6.14b. The high side is rated at I_2 without overloading any of the windings, and in this condition, from Kirchhoff's current law, the current on the low side is $(I_1 + I_2)$. The product of the high-side voltage and current ratings is (using the product on the low side will yield similar results) the autotransformer rating – that is, the volt-amperes that can be transferred through it:

$$\text{autotransformer rating} = (V_1 + V_2) I_2 \tag{6.24}$$

Comparing Equations 6.23 and 6.24, the ratings of the two arrangements are related as

$$\text{autotransformer rating} = \left(1 + \frac{V_1}{V_2}\right) \times \text{two-winding transformer rating} \tag{6.25}$$

Since the two-winding transformer rating equals the system VA transfer requirement, the needed autotransformer is equivalent to a two-winding transformer of the following rating:

$$\text{equivalent two-winding transformer rating}$$
$$= \text{transfer requirement} \div \left(1 + \frac{V_1}{V_2}\right) \tag{6.26}$$

From Figure 6.14b, expressing the term in the bracket in Equation 6.26 in terms of the high-side voltage $V_H (= V_1 + V_2)$ and the low-side voltage $V_L (= V_1)$,

$$\text{equivalent two-winding transformer rating}$$
$$= \left(1 - \frac{V_L}{V_H}\right) \times \text{transfer requirement} \tag{6.27}$$

If the high-side and low-side voltages are not far apart, the equivalent two-winding transformer rating of an autotransformer can be much smaller than the transfer requirement.

Example 6.2
In a system, 1 MVA has to be transferred with the low-side voltage of 22 kV and the high-side voltage of 33 kV [2, 3]. Calculate the equivalent two-winding transformer rating of an autotransformer to satisfy this requirement.

Solution   From Equation 6.27, the answer is 333 kVA.

The previous example shows that only a 333 kVA autotransformer will suffice, whereas we would otherwise require a 1,000 kVA conventional two-winding transformer. Therefore, an autotransformer is physically smaller and less costly. Efficiencies of autotransformers are greater than those of their two-winding counterparts, whereas their leakage reactances are comparatively smaller. The main disadvantage of autotransformers is that there is no electrical (galvanic) isolation between the two sides – however, it is not always needed. Hence, autotransformers are frequently used in power systems.

6.8 PHASE SHIFT INTRODUCED BY TRANSFORMERS

There are two primary ways that phase shifts in voltages are introduced by transformer connections in three-phase power systems. One common practice is to connect transformers in a Y on one side and a delta on the other, resulting in a phase shift of 30 degrees in voltages between the two sides. Another type of transformer connection is where a controllable phase shift is desired to regulate the power flow through the transmission line to which the transformer is connected. We will look at both of these cases.

6.8.1 Phase Shift in Delta-Y Transformers

Transformers connected in delta-Y, as shown in Figure 6.15a, result in a 30-degree phase shift that is shown by the phasor diagram in Figure 6.15b. To boost the voltages produced by the generators, the low-voltage sides are connected in a delta and the high-voltage sides are connected in a grounded Y. On the delta-connected side, the terminal voltages, although isolated from ground, can be visualized by hypothetically connecting very large but equal resistances from each terminal to a hypothetical neutral n. Thus, $\bar{V}_{An}$ is the terminal-A voltage with respect to the hypothetical neutral n. As shown in Figure 6.15b, $\bar{V}_{An}$ leads $\bar{V}_a$ by 30 degrees, and the magnitude of the two voltages can be related as follows:

$$\bar{V}_{AC} = \left(\frac{n_1}{n_2}\right)\bar{V}_a, \quad \text{and thus} \quad \bar{V}_{An} = \frac{1}{\sqrt{3}}\left(\frac{n_1}{n_2}\right)\bar{V}_a\, e^{j30^\circ} \qquad (6.28)$$

Based on Equation 6.28, the per-phase equivalent circuit is shown in Figure 6.15c.

6.8.2 Phase-Angle Control

As discussed in Chapter 2, power flow between two AC systems depends on sin δ, where δ is the difference in the phase angles between the two. Figure 6.16a shows a phase-regulating transformer arrangement for achieving a phase shift.

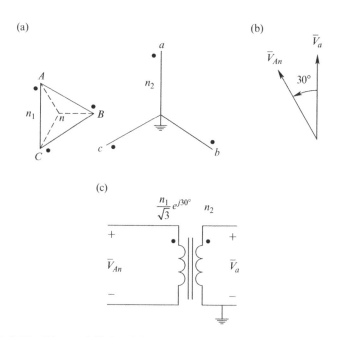

FIGURE 6.15 Phase shift in delta-Y connected transformers. (a) Winding arrangement; (b) phasor diagram; (c) equivalent circuit.

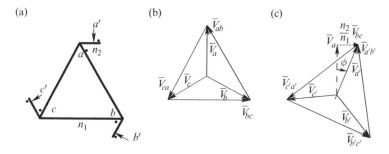

FIGURE 6.16 Transformer for phase-angle control. (a) Winding arrangement; (b) phasor diagram of incoming voltages a–b–c; (c) phasor diagram of outgoing voltages a′–b′–c′.

A phasor diagram of the incoming voltages $a - b - c$ is shown in Figure 6.16b and of the outgoing voltages $a' - b' - c'$ in Figure 6.16c, which shows that the outgoing voltages lag by an angle ϕ with respect to the incoming voltages. The sliders in Figure 6.16a represent finite taps.

6.9 THREE-WINDING TRANSFORMERS

Often, in power system transformers and autotransformers, an isolated third winding is added, to which are connected capacitors or power-electronics-based apparatus to supply reactive power to support the system voltage. The delta-connected tertiary winding usually has a low voltage rating and a much lower MVA rating. The purpose of the tertiary winding is to provide a path for zero-sequence currents to circulate, as discussed in Chapter 13, and to connect devices to supply reactive power.

The operating principle of three-winding transformers is an extension of the two-winding transformers discussed earlier. Often, an autotransformer arrangement is used between the high and low sides, and a tertiary winding is added and connected, as shown in Figure 6.17a. This can be represented as shown in Figure 6.17b.

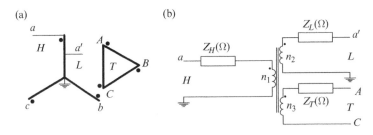

FIGURE 6.17 Three-winding autotransformer. (a) Winding arrangement; (b) circuit representation.

6.10   THREE-PHASE TRANSFORMERS

As discussed earlier, power transformers consist of three phases, where separate single-phase transformers may be connected in a Y or delta arrangement. In contrast, in a three-phase transformer, all the windings are placed on a common core, resulting in reduced core cost. The type of transformer to purchase depends on factors such as initial installed cost, maintenance costs, operating cost (efficiency), reliability, etc. Three-phase units have lower construction and maintenance costs and can be built to the same efficiency ratings as single-phase units. Many electrical systems have mobile substations and spare emergency transformers to provide backup in case of failure [4].

In extra-high-voltage systems at high power levels, single-phase transformers are used because the size of three-phase transformers at these levels makes their transportation from the manufacturing site to the installation site difficult. For unique requirements, it may be economical to have emergency spare single-phase transformers (one-third of the total rating) rather than spare three-phase transformers [4].

A three-phase transformer is either a shell or a core type. In balanced three-phase operation, the equivalent circuit of a three-phase transformer of either type, on a per-phase basis, is similar to that of a single-phase transformer. Only under unbalanced operation, such as during unsymmetrical faults, does the difference between these three-phase transformer types appear.

6.11 REPRESENTING TRANSFORMERS WITH OFF-NOMINAL TURNS RATIOS, TAPS, AND PHASE SHIFTS

As shown by Example 6.1, if a transformer with a nominal turns ratio and without phase shift is encountered, it can be simply represented by its leakage impedance. Special consideration must be given to transformers with off-nominal turns ratios, taps, and phase shifts. Figure 6.18 shows a general representation where the voltage transformation is by $1: t$, where t is a real number in case of an off-nominal turns ratio and a complex number in case of a phase shifter; Z_ℓ is the leakage impedance, and $Y_\ell = 1/Z_\ell$.

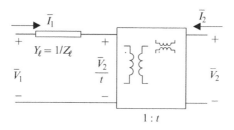

FIGURE 6.18 General representation of an autotransformer and a phase shifter.

Considering the leakage impedance separately, the voltage transformation across the ideal-transformer portion is as shown in Figure 6.18, and the sum of the complex powers into it must be equal to zero so that

$$\frac{\bar{V}_2}{t}\bar{I}_1^* = -\bar{V}_2\bar{I}_2^* \qquad (6.29)$$

In Figure 6.18,

$$\bar{I}_1 = \left(\bar{V}_1 - \frac{\bar{V}_2}{t}\right)Y_\ell \qquad (6.30)$$

From Equations 6.29 and 6.30, and recognizing that $t t^* = |t|^2$,

$$\bar{I}_2 = -\frac{\bar{I}_1}{t^*} = -\bar{V}_1\frac{Y_\ell}{t^*} + \bar{V}_2\frac{Y_\ell}{|t|^2} \qquad (6.31)$$

From Equations 6.30 and 6.31,

$$
\begin{bmatrix} \bar{I}_1 \\ \bar{I}_2 \end{bmatrix} = \begin{vmatrix} Y_\ell & -\dfrac{Y_\ell}{t} \\ -\dfrac{Y_\ell}{t^*} & \dfrac{Y_\ell}{|t|^2} \end{vmatrix} \begin{bmatrix} \bar{V}_1 \\ \bar{V}_2 \end{bmatrix}
\tag{6.32}
$$

This nodal representation is similar to the one discussed in Chapter 5 for representing the system for power-flow studies. It is discussed later, separately for autotransformers and phase shifters.

6.11.1 Off-Nominal Turns Ratios and Taps

The per-unit representation of a transformer with off-nominal turns ratio and taps (without phase shift) is by a transformer with a turns ratio $1: t$ where t is real in Figure 6.19a. Equation 6.32 can be represented by a pi-circuit, as shown in Figure 6.19b, where $t^* = t$ since t is real. The shunt admittances in Figure 6.19b can be calculated as shown.

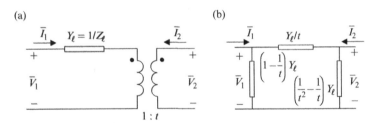

FIGURE 6.19 (a) Transformer with an off-nominal turns ratio or taps in per-unit (t is real); (b) pi-circuit representation.

Example 6.3
In the transformer in Figure 6.20a in per-unit, the leakage impedance $Z_\ell = j0.1$ pu. The turns ratio $t = 1.1$ in per-unit to boost the voltage at side 2. Calculate the parameters in the pi-circuit in Figure 6.20b for use in power-flow programs.

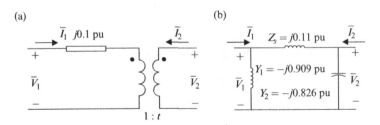

FIGURE 6.20 (a) Transformer with a turns-ratio for Example 6.3; (b) pi-circuit representation.

Solution   In this example, $Y_\ell = 1/Z_\ell = -j10$ pu . In the pi-circuit, $Y_1 = -j0.909$ pu, where the susceptance value is negative, implying that it is inductive. $Y_2 = j0.826$ pu, where the susceptance value is positive, implying that it is capacitive. The corresponding pi-circuit model is shown in Figure 6.20b, which clearly shows how the LC circuit results in greater magnitude V_2 than V_1.

6.11.2   Representing Transformer Phase Shift

In a phase-shift transformer arrangement, t in the Y-matrix in Equation 6.32 is complex. Therefore, $Y_{12} \neq Y_{21}$. Consequently, a transformer with a complex turns ratio cannot be represented by a pi-circuit; rather, it calls for an admittance matrix from Equation 6.32 to be used in the power-flow studies.

Two Winding Transformer Data Record   ×

Power Flow   Short Circuit

Line Data

From Bus Number	1	From Bus Name BUS 1 345.00	☑ In Service
To Bus Number	4	To Bus Name BUS 4 345.00	☑ Metered on From end
Branch ID	1	Transformer Name AUTO XMER	☑ Winding 1 on From end
		Vector Group Ya0 ...	

I/O Data

Winding I/O Code	Impedance I/O Code	Admittance I/O Code
1 - Turns ratio (pu on bus base kV) ∨	1 - Z pu (winding kV system MVA) ∨	1 - Y pu (system base) ∨

Transformer Impedance Data

		Transformer Nominal Ratings Data		
Specified R (pu)	Specified X (pu)	Winding 1 Ratio (pu)	Winding 1 Nominal kV	Ratings (MVA)
0.000000	0.020000	0.95000	345.00000	RATE1 0.0
Magnetizing G (pu)	Magnetizing B (pu)	Winding 2 Ratio (pu)	Winding 2 Nominal kV	RATE2 0.0
0.00000	0.00000	1.00000	345.00000	RATE3 0.0
Impedance Table		Winding (1-2) Angle (degrees)	Winding MVA	RATE4 0.0
0		0.00	100.0000	RATE5
R table corrected (pu)	X table corrected (pu)	Control Data		
0.00000	0.02000	Controlled Bus Number	Controlled Bus Name	Control Mode
		0		0- None ∨

Owner Data

Owner		Fraction	☐ Controlled Bus On Winding Side	☑ Auto Adjust	Load Drop Comp
1	Select ...	1.000	Tap Positions	Wnd Connect Angle	Load Drop Comp R (pu)
0	Select ...	1.000	33	0.00000	0.00000
0	Select ...	1.000	R1max (pu)	R1min (pu)	Load Drop Comp X (pu)
0	Select ...	1.000	1.10000	0.90000	0.00000
			Vmax (pu)	Vmin (pu)	
			1.10000	0.90000	

FIGURE 6.21   Two-winding transformer data.

6.12   TRANSFORMER MODEL IN PSS®E

PSS®E offers various attributes to define transformers in the power flow. It has options to specify whether it's a core type or an autotransformer. There is a choice of star, star-grounded, and delta-winding definitions. It can be a two-winding or three-winding transformer. It can be a tap-controlled or phase-shift-controlled transformer or neither. We can also model the impedance of the phase-shifting transformer to vary as a function of its phase-shift angle; this is defined by piecewise linear tables known as *impedance correction tables*. This is necessary because, as seen in Section 6.11.2, the Y-bus matrix is not symmetric, and hence the admittance from the primary bus to the secondary bus is not equal. This can affect the power-flow results if ignored. The steady-state two-winding transformer data options are typically as shown in Figures 6.21 and 6.22.

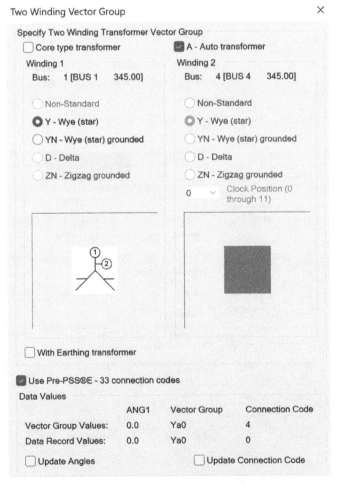

FIGURE 6.22   Two-winding vector group definitions.

REFERENCES

1. P. Kundur. 1994. *Power System Stability and Control.* McGraw Hill.
2. Central Station Engineers. 1950. *Electrical Transmission and Distribution Reference Book.* Westinghouse Electric Corporation.
3. P. Anderson. 1995. *Analysis of Faulted Power Systems.* Wiley-IEEE Press.
4. U.S. Department of Agriculture, Rural Utilities Service. *Design Guide for Rural Substations.* RUS Bulletin 1724E-300. https://www.rd.usda.gov/files/UEP_Bulletin_1724E-300.pdf.

PROBLEMS

6.1 A transformer is designed to step down the applied voltage of 2400 V (rms) to 240 V (rms) at 60 Hz. Calculate the maximum rms voltage that can be applied to the high side of this transformer without exceeding the rated flux density in the core if this transformer is supplied by a frequency of 50 Hz.

6.2 Assume the transformer in Figure 6.4a is ideal. Winding 1 is applied a sinusoidal voltage in a steady state with $\overline{V}_1 = 120 \text{ V} \angle 0°$ at a frequency $f = 60$ Hz. $N_1/N_2 = 3$. The load on winding 2 is a series combination of R and L with $Z_L = (5 + j\,3)\ \Omega$. Calculate the current drawn from the voltage source.

6.3 Consider an ideal transformer, neglecting the winding resistances, leakage inductances, and core loss. $N_1/N_2 = 3$. For a voltage of 120 V (rms) at a frequency of 60 Hz applied to winding 1, the magnetizing current is 1.0 A (rms). If a load of 1.1 Ω at a power factor of 0.866 (lagging) is connected to the secondary winding, calculate $\overline{I}_1$.

6.4 A 2400/240 V, 60 Hz transformer has the following parameters in the equivalent circuit in Figure 6.5: the high-side leakage impedance is (1.2 + j 2.0) Ω, the low-side leakage impedance is (0.012 + j 0.02) Ω, and X_m at the high side is 1800 Ω. Neglect R_{he}. Calculate the input voltage if the output voltage is 240 V (rms) and supplying a load of 1.5 Ω at a power factor of 0.9 (lagging).

6.5 Calculate the equivalent-circuit parameters of a transformer if the following open-circuit and short-circuit test data is given for a 60 Hz, 50 kVA, 2400:240 V distribution transformer:

open-circuit test with high side open : $Voc = 240$ V, $Ioc = 5.0$ A, $Poc = 400$ W

short-circuit test with low side shorted : $Vsc = 90$ V, $Isc = 20$ A, $Psc = 700$ W

6.6 Three two-winding transformers are grounded Y-delta connected to 230 kV on the Y side and 34.5 kV on the delta side. The combined three-phase rating of these transformers is 200 MVA. The per-unit reactance is 11% based on the transformer rating. Calculate the rated values of voltage and currents, and X_{ps} and X_{sp} (in Ω) in Figures 6.8a and 6.8b.

6.7   In a transformer, the leakage reactance is 9% based on its rating. What will its percent value be if, simultaneously, the voltage base is doubled and the MVA base is halved?

6.8   Efficiency in a power transformer is 98.6% when loaded such that its core losses equal the copper losses. Calculate the per-unit resistance of this transformer.

6.9   Using Figure 6.9, calculate $\overline{V}_p$ in per-unit, if $\overline{V}_s(\mathrm{pu}) = 1\angle 0°$, rated $\overline{I}(\mathrm{pu}) = 1\angle -30°$, and $X_{\mathrm{tr}} = 11\%$. Draw the relationship between these calculated variables in a phasor diagram.

6.10   The % regulation for a transformer is defined as

$$\% \text{ regulation} = 100 \times \frac{V_{\text{no-load}} - V_{\text{rated}}}{V_{\text{rated}}}$$

where V_{rated} is the voltage at the rated kVA loading at the specified power load. Calculate the % regulation for the transformer in Problem 6.9.

6.11   Calculate the regulation in Problem 6.10 if the rated $\overline{I}(\mathrm{pu}) = 1\angle 0°$. Compare this value with that in Problem 6.10.

6.12   Derive the expression for the full-load regulation, where θ is the power-factor angle, which is taken as positive when the current lags the voltage, and X_{tr} is in per-unit.

6.13   In Example 6.2, the autotransformer is loaded to its rated MVA at unity power factor. Calculate all voltages and currents in the autotransformer in Figure 6.14b.

6.14   In Example 6.2, if a two-winding transformer is selected, its efficiency is 99.1%. If the same transformer is connected as an autotransformer, calculate its efficiency based on the MVA rating it is capable of. Assume a unity power factor of operation in both cases.

6.15   In the Y-delta connected transformer shown in Figure 6.15a, show all the phase and line-line voltages on both sides of the transformer, similar to Figure 6.15b. The line-line voltage on the Y side is 230 kV and on the delta side is 34.5 kV. Assume the phase angle $\overline{V}_a$ is equal to 90 degrees.

6.16   In Problem 6.17, the Y-connected load on the delta side is purely resistive, and the load-current amplitude is 1 pu. Draw the currents within the delta windings and those drawn from the Y side in per-unit.

INCLUDING TRANSFORMERS IN POWER-FLOW STUDIES USING PSS®E

6.17   The power flow in a three-bus example power system is calculated using MATLAB in Example 5.4. Repeat this using PSS®E, including an autotransformer in this example for voltage regulation, as described on the accompanying website, and calculate the results of the power-flow study.

6.18 The power flow in a three-bus example power system is calculated using MATLAB in Example 5.4. Repeat this using PSS®E, including a phase-shifting transformer in this example for controlling power flow, as described on the accompanying website, and calculate the results of the power-flow study.

7

GRID INTEGRATION OF INVERTER-BASED RESOURCES (IBRS) AND HVDC SYSTEMS

OBJECTIVES

This chapter provides an understanding of how large-scale renewable energy power plants, also called inverter-based resources (IBRs), distributed energy resources (DERs), and battery energy storage systems (BESSs), are integrated into the utility grid. This chapter also demonstrates how these can be modeled in PSS®E.

Renewable resources are often remotely located, so this chapter also discusses DC transmission lines as a possible alternative to AC transmission lines. Two types of high-voltage DC (HVDC) transmission systems are described:

- HVDC-VSC using voltage-source converters (VSCs)
- HVDC-LCC using thyristor-based line-commutated converters (LCCs) for very large power transfer over long distances

An HVDC-VSC system is modeled in PSS®E.

This chapter also briefly describes the IEEE P2800 standard being considered for interfacing IBRs to the utility grid.

Electric Power Systems with Renewables: Simulations Using PSS®E, Second Edition. Ned Mohan and Swaroop Guggilam.
© 2023 John Wiley & Sons, Inc. Published 2023 by John Wiley & Sons, Inc.
Companion Website: www.wiley.com/go/mohaneps

7.1 CLIMATE CRISIS

The climate crisis is upon us, and leading climate scientists worldwide are warning that it will get much worse, with catastrophic consequences, if we fail to act now to drastically curb greenhouse gas emissions from using fossil fuels. Technology alone cannot stop global warming and climate change; for that to happen, human attitudes must also change. However, technology has some hopeful solutions, primarily using renewable energy to replace our use of fossil fuels for electric power generation and transportation.

Various agencies report that utility-scale renewables such as wind and solar are cheaper than their fossil-fuel counterparts. The cost of combining these variable energy resources with battery storage is also declining rapidly. Consequently, almost all new construction of power plants in the United States is either solar or wind.

7.2 INTERFACE BETWEEN RENEWABLES/BATTERIES AND THE UTILITY GRID

As discussed in Chapter 3, large-scale wind and photovoltaic (PV) systems require a power-electronics interface. Similarly, battery-storage systems, where the battery voltage is DC, need a power electronics interface to integrate them into the grid. These are called inverter-based resources (IBRs). The schematic in Figure 7.1 shows the interface to the grid for PVs, wind generators, and storage batteries, respectively.

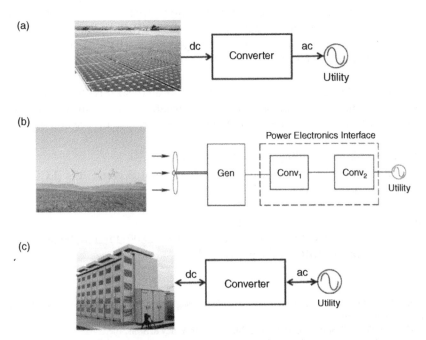

FIGURE 7.1 Schematic of interfacing IBRs with the grid: (a) PV, (b) wind, and (c) battery storage.

In discussing these power-electronics-based interfaces, the term *converter* is more generic, whereas the term *inverter* implies an interface where power generally flows from the DC side to the AC side. However, these terms may be used interchangeably.

7.2.1 Voltage-Source Converters (VSCs)

The role of these converters is shown in Figure 7.1 in a block-diagram form for various applications. The figure makes it clear that in a VSC, there is DC voltage on one side and three-phase AC voltage on the other side. The sinusoidal three-phase AC voltage may be at the line frequency with a fixed amplitude or variable frequency (e.g., between 6 and 60 Hz) and with variable amplitude, as is the case in wind turbines. In the case of battery storage systems, the power through this interface is bidirectional, depending on the charge or discharge mode.

VSCs utilize the concepts of switch-mode power electronics and consist of semiconductor devices that are operated as bipositional switches: that is, they are either on or off. They are described in **Appendix 7A**. However, as a brief introduction, using bipositional switches operating at very high frequencies (several kilohertz to several hundreds of kilohertz), a voltage of any form (amplitude and frequency) is chopped into discrete quantities. Using an approach called pulse-width modulation (PWM), this chopped waveform is resynthesized. Here, it will suffice to represent VSCs with the block diagram shown in Figure 7.2, where the voltage on the DC side, across the capacitor, is V_{dc} and the three-phase sinusoidal voltage on the AC side is of a frequency f and line-line (rms) magnitude, V_{LL}. Through these converters, the power flow is inherently bidirectional.

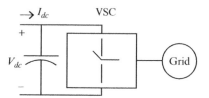

FIGURE 7.2 Voltage-source converter (VSC).

The relationship between the DC-side voltage and the AC-side voltage V_{LL} (line-to-line, rms) at the converter terminal is

$$V_{dc} = k_{PWM} V_{LL} \tag{7.1a}$$

where k_{PWM} is the conversion coefficient, which can be controlled to be in a range between 0 and $\sqrt{2}$:

$$0 \leq k_{PWM} \leq \sqrt{2} \tag{7.1b}$$

Note that the energy efficiencies of these converters are very high, greater than 90 to 95%; therefore, for our discussion, we can assume them to be ideal, with 100% efficiency. This implies that the three-phase power equals the power on the DC side:

$$P = V_{dc}I_{dc} = \sqrt{3}V_{LL}I_L \cos\phi \tag{7.2}$$

ϕ is the power-factor angle between the phase voltages and phase currents on the AC side of this converter, as shown in Figure 7.3. Here, $\overline{V}_{conv}$ is the rms phase voltage of magnitude $V_{LL} / \sqrt{3}$ at the converter terminals and is taken as the reference phasor. With the current direction out of the VSC in Figure 7.2, Figure 7.3 shows the power being sourced by the converter.

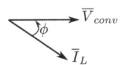

FIGURE 7.3   Voltage and current relationship on the AC side at the converter terminals.

In addition to the real power, the three-phase reactive power on the AC side, supplied by the VSC, is

$$Q = \sqrt{3}V_{LL}I_L \sin\phi \tag{7.3}$$

The direction and the amount of real and reactive powers depend on the relationship between the VSC voltage and the rest of the system connected on the AC side in accordance with the concepts discussed in Section 2.5 in Chapter 2. This is represented in Figure 7.4 on a per-phase basis, where the AC system voltage, $\overline{V}_{grid}$, is taken as the reference phasor, $V_{grid}\angle 0°$.

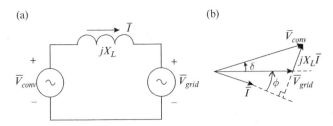

FIGURE 7.4   Power transfer between the VSC and the AC (grid) system. (a) VSC connected to the grid; (b) phasor diagram.

All the quantities are at the same frequency in Figure 7.4a, and the current is

$$\bar{I} = \frac{\bar{V}_{conv} - \bar{V}_{grid}}{jX_L} \tag{7.4}$$

At the grid terminals, the complex three-phase power can be written as

$$S_{grid} = P_{grid} + jQ_{grid} = 3(V_{grid}\bar{I}^*) \tag{7.5}$$

Therefore, from Equation 2.44 derived in Chapter 2, on a three-phase basis,

$$P_{grid} = 3\left(\frac{V_{conv}V_{grid}}{X_L}\right)\sin\delta \tag{7.6}$$

and

$$Q_{grid} = 3\left(\frac{V_{conv}V_{grid}\cos\delta - V_{grid}^2}{X_L}\right) \tag{7.7}$$

Here, the real power is the same at the converter terminals as at the grid terminals, provided there is no loss associated with the inductance. If the real power is zero, then, from Equation 7.6, $\sin\delta$ and the angle δ are equal to zero. Under this condition, from Equation 7.7,

$$Q_{grid} = \frac{V_{grid}}{X_L}\left(V_{conv} - V_{grid}\right) \quad \text{(if } P_{grid} = 0\text{)} \tag{7.8}$$

which shows that the reactive power transfer from the converter to the grid depends on the difference $(V_{conv} - V_{grid})$ between the two voltage magnitudes. However, the real power transfer doesn't have to be zero, and it could be in any direction, as dictated by Equation 7.6; and the reactive power can be in either

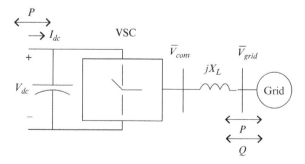

FIGURE 7.5 Real and reactive power through a VSC and the grid.

direction, dictated by Equation 7.7. This is shown graphically in Figure 7.5 with a one-line diagram.

This is a unique capability of VSCs to supply or absorb reactive power, independent of the direction of the power flow through the converter. This capability is used in HVDC-VSC, as we will see later in this chapter. This is also the principle behind static synchronous compensators (STATCOMs), where $P = 0$, and it can supply reactive power Q, as a shunt-connected capacitor would, or absorb Q as a shunt-connected inductor would.

7.2.1.1 VSCs in Wind Generators

The discussion so far assumes a VSC connected to the grid, often through a transformer. However, this discussion applies equally well if the AC side of the VSC is a wind generator producing voltages that vary in frequency and amplitude depending on the wind speed. This is shown in Figure 7.6.

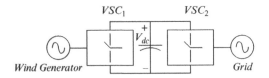

FIGURE 7.6   A VSC system in a wind generator.

7.2.2   Modeling Voltage-Source Converters (VSCs) in PSS®E

For a given grid voltage, the real and reactive power of a VSC can be controlled by synthesizing the VSC output voltage of an appropriate phase-angle δ and the voltage magnitude in accordance with Equations 7.6 and 7.7. This allows us to model it as a generator for the power-flow analysis.

A great deal of state-of-the-art research is happening regarding steady-state and dynamic models of inverter-based resources. We briefly discuss here one of the common models: the *WECC model* [1], which is used to represent IBRs in a steady-state power-flow analysis. In general, IBRs can be spread across the geographical area and may contain multiple segments of connections before the output power reaches the high-voltage transmission system. The initial layer can consist of hundreds of wind plants and solar farms, each connected through a step-up transformer to a feeder network. The feeder network then connects to local substations, the collector system. Finally, a step-up station transformer is connected directly to the high-voltage transmission system or via an interconnection transmission line. This contact point is also known as the *point of interconnection*. Additionally, reactive power-compensation devices may be needed in the intermediate transition depending on the interconnection requirements. It may not be sufficient to represent all this as a single generator on the high-voltage bus. The WECC model is shown in Figure 7.7.

In PSS®E, the equivalent generator can be modeled by using the generator model and specifying the wind control mode. Note that this is still an

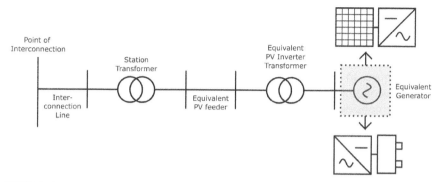

FIGURE 7.7   Steady-state modeling of a voltage-source converter in PSS®E. Source: [1].

approximate equivalent model, and renewable resources may operate very close to its limits. This model is also sufficient for positive sequence representation in transient stability analysis for bulk power systems.

The technical challenge lies in parametrizing these individual components [2]. We must ask ourselves how we can calculate the resistance (R) and reactance (X) values of the equivalent PV feeder line or the collector system. How to determine the equivalent PV inverter transformer parameters is an opportunity for ongoing and future research.

Example 7.1

In Example 5.4, replace the generator on bus 2 with an equivalent IBR representation using the WECC model. Perform a power-flow analysis, and compare the results.

Solution: The parameters for the individual components of the WECC model, the method to model these components in PSS®E, and the PSS®E simulation files for this example are included on the accompanying website. A sample representation of the resulting model is shown in Figure 7.8.

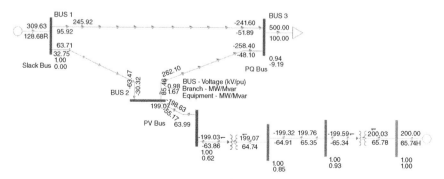

FIGURE 7.8   Sample WECC representation of the generator as an inverter-based resource in the three-bus example.

7.3 HIGH-VOLTAGE DC (HVDC) TRANSMISSION SYSTEMS

For transmitting large amounts of power over long distances, such as from a remote source to load centers far away, a high-voltage DC (HVDC) system can be more economical than an AC transmission system. Thus an HVDC system should be considered for such purposes. Repurposing existing AC transmission lines to HVDC should also be considered. For underwater transmission through cables, HVDC is almost a necessity. Lately, for stability reasons, back-to-back HVDC systems (without any transmission lines) are being built to transfer power between two AC systems where an AC link will not be stable. More such back-to-back systems are expected to be used in the future.

As mentioned earlier, there are two types of HVDC transmission systems:

- HVDC-VSC using voltage-source converters (VSCs)
- HVDC-LCC using thyristor-based line-commutated converters (LCCs) for very large power transfer over long distances

7.3.1 HVDC-VSC Systems

To transfer power from renewable energy sources to where it is required, HVDC-VSC systems can be considered. This is also the case where power levels are not too high (less than GW) and the transmission distances are not very long. Otherwise, thyristor-based HVDC-LCC systems are used.

An HVDC-VSC system is shown in Figure 7.9. In such a system, the direction of power flow can be reversed by changing the direction of the current in the DC line. The figure shows a one-line diagram of one such system in operation in the United States that operates at $+/-150$ kV and is rated at 330 MW [3].

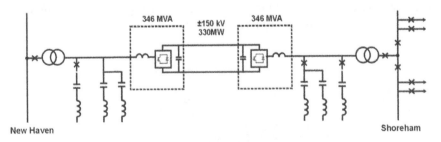

FIGURE 7.9 Block diagram of a voltage-link HVDC system [3].

In the system shown in Figure 7.9, each converter can be independently controlled and can either absorb or supply reactive power in a controllable manner. There is a capacitor on the DC side of each converter in parallel that appears as a voltage port on the DC side, and hence the converters in such a system are VSCs.

Each of these converters is as described earlier, to transfer power between the AC and DC sides and supply/absorb reactive power on the AC side.

The system from Figure 7.9 is redrawn in Figure 7.10, connecting two systems with a high-voltage DC transmission line of resistance R_d and inductance L_d.

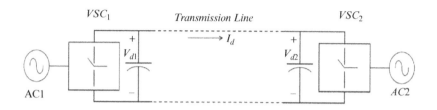

FIGURE 7.10 HVDC-VSC with a DC transmission line of resistance R_d and inductance L_d.

Here,

$$I_d = \frac{V_{d1} - V_{d2}}{R_d} \qquad (7.9)$$

Therefore, controlling V_{d1}, for example, where V_{d2} is kept constant, the current I_d and hence the power transmitted through the transmission line can be controlled.

7.3.1.1 HVDC VSC Modeling in PSS®E

The VSC two-terminal DC transmission line offers grid flexibility by allowing control of the voltage and active and reactive power. It can be used to simulate either a point-to-point system or a back-to-back system. The steady-state power-flow model representation of VSC DC can be summarized using Figure 7.11 and Table 7.1.

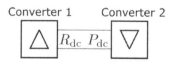

FIGURE 7.11 VSC DC model in PSS®E.

TABLE 7.1 VSC DC control modes

DC control type	AC control mode
Voltage control	Voltage control
MW flow control	Power-factor control

Each converter has the control options defined in Table 7.1. Typically, the DC control modes for both converters are such that one should operate at voltage control mode and the other at MW flow control. If both the converter's DC control modes are the same, no power transfer can happen, and the DC line is ignored for the steady-state power-flow analysis. AC control modes can be dependent on the application. Both converters can be set to control their respective AC side voltages, or one converter can be used to control its AC side voltage and the other converter can be used to maintain a constant power factor.

Example 7.2
In the example system in Figure 5.4, replace the transmission line bus 2 to bus 3 with a VSC DC line. Bus 2's converter operates at DC (345 kV) and AC (1.05 pu) voltage control modes. Bus 3's converter operates at DC MW flow (264.2 MW) and AC (0.98 pu) voltage control modes. The DC line resistance is 3.47 ohms.

Solution   This allows mimicking the effect of AC transmission line. The graphical result from PSS®E is shown in Figure 7.12, and the files related to this example are included on the accompanying website.

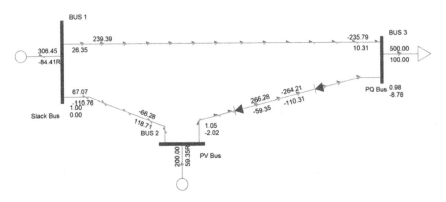

FIGURE 7.12   Power-flow results for Example 7.2.

7.3.2   HVDC-LCC Transmission Systems

At very high power levels and over long distances, the choice remains thyristor-based systems. These systems are explained in **Appendix 7B**. An example is a system that transmits 3,800 MW at ±500 kV over a distance of 1,361 km in the western part of the United States. At very high power levels (more than 1,000 MW), the use of thyristors, at least for now, represents a reasonable choice.

Figure 7.13 shows the block diagram of an HVDC-LCC transmission system. Assuming the power flows from left to right, the voltages in AC system 1 at the sending end are stepped up using transformers. These voltages are rectified into DC using a thyristor-based converter, where the AC line voltages provide the commutation of current within the converter: hence the name line-commutated converter (LCC). The three-phase AC currents drawn from system AC1 turn into a DC current on the other side of the converter. This current is transmitted over the DC line, where additional series inductance is added to ensure that the DC current is smooth and as free of ripples as possible. At the receiving end, a similar thyristor-based converter converts the DC current into three-phase AC currents and injects them into AC system AC2 through step-down transformers. In Figure 7.13, R_d is the resistance, and inductance L_d is the inductance of the DC transmission line.

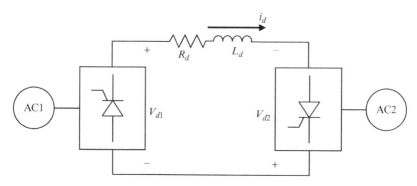

FIGURE 7.13   Block diagram of an HVDC-LCC system.

The DC voltage across each converter, V_{d1} and V_{d2}, defined with the polarities shown, can be controlled by controlling their delay angles α_1 and α_2 in a range of 0 degrees to approximately 160 degrees, as discussed in Appendix 7B.

By controlling the average DC voltages at the two ends of the transmission line, the average current and the power flow through the transmission line can be controlled. The DC current can be expressed as

$$I_d = \frac{V_{d1} + V_{d2}}{R_d} \tag{7.10}$$

where the + sign is needed due to the polarity with which v_{d2} is defined. Since the line resistance R_d is generally very small, V_{d1} and V_{d2} are very close in magnitude and opposite in value.

7.3.2.1   Control of Power Flow

For the power flow from system 1 to system 2, as shown in Figure 7.13, V_{d2} is made negative (with the polarity shown) by controlling α_2 such that converter 2 operates as an inverter and establishes the voltage of the DC line. Converter 1 is operated as a rectifier, with a positive value of V_{d1} at a delay angle α_1, such that it is greater than V_{d2}, and it controls the current in the DC line. The converse is true for these two converter voltages if the power is to flow from system 2 to system 1.

In a converter such as the one on the right, operating in an inverter mode, the delay angle α_2 is controlled such that the inverter can perform its function with a constant margin of safety called the *extinction angle* γ.

Therefore, with converter 2 operating as an inverter at a constant minimum extinction angle $\gamma_{\min}$, the voltage V_{d1} given by Equation 7.10 is plotted in Figure 7.14 as a function of the DC-link current I_d. Converter 1 is operating as a rectifier with its delay angle controlled to maintain the DC-link current at its reference value $I_{d,ref}$. Therefore, its characteristic appears as a vertical line in Figure 7.14. The intersection of the inverter and the rectifier characteristics establishes the operating point in terms of the voltage and current in this HVDC system.

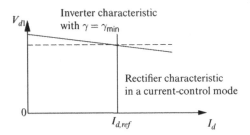

FIGURE 7.14   Control of an HVDC system [4].

7.4 IEEE P2800 STANDARD FOR INTERCONNECTION AND INTEROPERABILITY OF INVERTER-BASED RESOURCES INTERCONNECTING WITH ASSOCIATED TRANSMISSION ELECTRIC POWER SYSTEMS

There is a need for specifications and standards to address the performance requirements of inverter-based resources as their penetration levels increase. IEEE P2800 is such a standard being considered [5]. At the time of writing this textbook, it is still being voted upon to help planners, manufacturers, developers, and power grid operators improve power grid performance.

The intent of this standard is to provide uniform technical minimum requirements for interconnecting IBRs to transmission and subtransmission systems for a reliable interconnection of IBRs to the bulk power system for voltage and frequency ride-through and active and reactive power control under normal as well as abnormal conditions. This standard will also apply to HVDC-VSCs, discussed earlier.

REFERENCES

1. Western Electricity Coordinating Council (WECC). https://www.wecc.org/Reliability/Solar%20PV%20Plant%20Modeling%20and%20Validation%20Guidline.pdf.
2. North American Electric Reliability Corporation (NERC). https://www.nerc.com/comm/PC/Pages/Power-Plant-Modeling-and-Verification-Task-Force-.aspx.
3. ABB Corporation. www.abb.com.
4. N. Mohan, T. Undeland, and W. P. Robbins. 2003. *Power Electronics: Converters, Applications, and Design*, 3rd ed. John Wiley & Sons.
5. IEEE. P2800—Standard for Interconnection and Interoperability of Inverter-Based Resources Interconnecting with Associated Transmission Electric Power Systems.

PROBLEMS

7.1 In Example 7.1, add another equivalent generator and a PV inverter transformer at the feeder bus, and split the total active power between the two generators. Perform power-flow analysis, and compare the results. The parameters of the model and the approach are described on the accompanying website.

7.2 In Example 7.2, modify one of the AC control modes to maintain a constant power factor, and analyze the power-flow results as described on the accompanying website in the three-bus example power system.

APPENDIX 7A OPERATION OF VOLTAGE SOURCE CONVERTERS (VSCS) [7A1]

VSCs consist of semiconductor devices such as transistors (MOSFETs and IGBTs) and diodes. In these converters, transistors operate as switches: they are on or off. There have been significant improvements in the ratings of these transistors so that they can withstand voltages in kilovolts when off and can carry currents in kiloamps when on. With continuing advancements in wide-bandgap semiconductors such as GaN and SiC, these devices have very low losses in their on state and similarly very low losses when they switch from one state to another. This allows them to be operated at very high frequency, with advantages such as reducing the size of filters to remove the switching ripple. The same is true of diodes.

To synthesize these three-phase voltages, let us assume that we hypothetically have three ideal transformers available with continuously variable turns ratios, as shown in Figure 7A.1a. We will soon see that these ideal transformers are functional representations of the switch-mode converter that is required. We will focus on only one of the three phases, as shown in Figure 7A.1b; the others are identical in functionality, where from practical considerations, d_a is restricted to a range $0 \leq d_a \leq 1$. With this restriction, v_{aN} cannot become negative, and therefore a DC offset of half the DC-bus voltage, $0.5V_d$, is introduced so that around this offset, the desired output voltage v_a, with respect to the output neutral, can become both positive and negative in a sinusoidal manner.

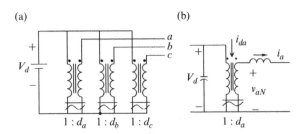

FIGURE 7A.1 Synthesis of sinusoidal voltages. (a) VSCs with ideal transformer representation; (b) phase *a* representation.

Therefore,

$$v_{aN} = 0.5V_d + \underbrace{\hat{V}_a \sin \omega t}_{v_a} \qquad (7A.1)$$

as shown in Figure 7A.2. In the ideal transformer in Figure 7A.1b, $v_{aN} = d_a V_d$ and hence the voltage in Equation 7A.1 can be obtained by varying the turns ratio $1:d_a$ with time, as follows, as shown in Figure 7A.2

$$d_a = 0.5 + \hat{d}_a \sin \omega t \qquad (7A.2)$$

where

$$\hat{V}_a = \hat{d}_a V_d \qquad (7A.3)$$

In Figure 7A.3a, all three-phase output voltages are shown; the waveforms are plotted in Figure 7A.3b. In these outputs, the DC offsets (actually,

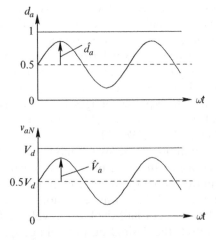

FIGURE 7A.2 Sinusoidal variation of the turns ratio d_a.

these common-mode voltages need not be DC as long as they are the same in all three phases) are canceled out from line-to-line voltages and hence can be ignored in consideration of the output voltages. Note that in Figure 7A.3b, by introducing the common-mode voltages of $V_d / 2$ in series with each phase, the maximum AC voltage magnitude is $\hat{V}_a = V_d / 2$. Using the so-called *space-vector PWM* (SV-PWM), as explained in [3], it is possible to modulate the common-mode voltage rather than keep it to $V_d / 2$ and thus get the line-line voltage peak to equal V_d at the limit. This increases the output voltage capability of such a converter by approximately 15%.

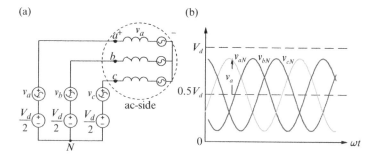

FIGURE 7A.3   (a) Three-phase voltages in VSC; (b) voltage waveforms.

Next, we will study how this ideal transformer functionality is obtained. As shown in Figure 7A.4a, a bipositional switch within a two-port circuit, a voltage-port on the DC side, and a current-port on the AC side, is used. This switch can be considered ideal, either up or down, for a switching signal q_a of 1 or 0, respectively. Such a switch can be constructed as shown in Figure 7A.4b using two diodes and two IGBTs that are provided complementary gate signals, q_a and q_a^-.

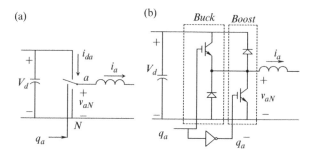

FIGURE 7A.4   Realization of the ideal transformer functionality. (a) Bipositional switch; (b) implementation.

In this bipositional switch, when $q_a = 1$ and therefore the upper IGBT is on and the lower one is off, the current through the output inductor can flow in either direction, through the upper IGBT if positive or through the upper diode if negative. In either of these cases, the potential of point a is the same as that of the upper DC bus and $v_{aN} = V_d$. Similarly, $q_a = 0$ results in $q_a^- = 1$, the output current will flow through either the bottom diode or the bottom IGBT, and hence $v_{aN} = 0$. Thus, the switching function q_a makes the switch operate as a bipositional switch, either up or down, regardless of the current direction.

PWM. We will operate this switch at a high frequency, two or three orders of magnitude higher than the fundamental frequency to be synthesized. For example, in synthesizing 60 Hz, the switching frequency f_s may be 6 kHz, which is 100 times higher, or even much higher. This results in the output waveform shown in Figure 7A.5a over one switching cycle $T_s (= 1/f_s)$, where the switching frequency f_s is kept constant. This output voltage has an average value, averaged over a switching-frequency time period T_s, that can be written with a "–" on top, as

$$\bar{v}_{aN} = d_a V_d \tag{7A.4}$$

where

$$d_a = \frac{T_{up}}{T_s} \tag{7A.5}$$

In Equations 7A.4 and 7A.5, T_{up} and d_a can be continuously varied with time, sinusoidally with a DC offset, as discussed earlier, so that the average voltage

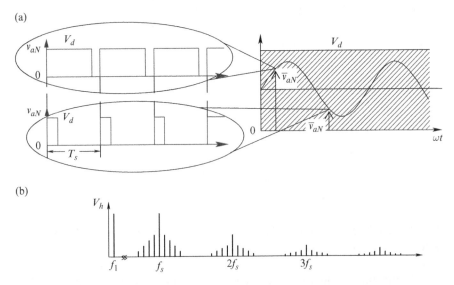

FIGURE 7A.5 PWM to synthesize a sinusoidal waveform. (a) PWM signal; (b) average voltage with harmonics.

given by Equation 7A.4 is the same as that obtained by the ideal transformers in Figure 7A.3, where v_{aN} and so on are ideal, without any ripple.

It should be recognized that in the output voltage, in addition to the desired average value Equation 7.22, there are unwanted switching harmonics, as shown in Figure 7A.5b, at and around the multiples of the switching frequency; these appear as side bands at and around harmonic h, where

$$h = k_1 f_s \pm k_2 f_0 \qquad (7A.6)$$

The design of converters must take these harmonics into account and provide adequate filtering so that they do not impact the AC system to which this converter is connected.

REFERENCES

7A1. N. Mohan. 2011. *Power Electronics – A First Course.* Wiley & Sons.

PROBLEMS

1. Plot $d_a(t)$ if the output voltage of the converter pole a is
   $$\bar{v}_{aN}(t) = \frac{V_d}{2} + 0.85\frac{V_d}{2}\sin(\omega_1 t), \text{ where } \omega_1 = 2\pi \times 60\,\text{rad/s}.$$

2. In a three-phase DC-AC inverter, $V_d = 350\,\text{V}$, $\hat{V}_{\text{tri}} = 1\,\text{V}$, the maximum value of the control voltage reaches $0.8\,\text{V}$, and $f_1 = 45\,\text{Hz}$. Calculate and plot (a) the duty ratios $d_a(t)$, $d_b(t)$, and $d_c(t)$; (b) the pole output voltages $\bar{v}_{aN}(t)$, $\bar{v}_{bN}(t)$, and $\bar{v}_{cN}(t)$; and (c) the phase voltages $\bar{v}_{an}(t)$, $\bar{v}_{bn}(t)$, and $\bar{v}_{cn}(t)$.

3. In a balanced three-phase DC-AC converter, the phase a average output voltage is $\bar{v}_{an}(t) = 112.5\sin(\omega_1 t)$, where $V_d = 300\,\text{V}$ and $\omega_1 = 2\pi \times 45\,\text{rad/s}$. The inductance L in each phase is 5 mH. The AC motor internal voltage in phase A can be represented as $e_a(t) = 106.14\sin(\omega_1 t - 6.6°)\,\text{V}$. (a) Calculate and plot $d_a(t)$, $d_b(t)$, and $d_c(t)$, and (b) sketch $\bar{i}_a(t)$ and $\bar{i}_{da}(t)$.

APPENDIX 7B OPERATION OF THYRISTOR-BASED LINE-COMMUTATED CONVERTERS (LCCS)

The block diagram of an HVDC-LCC system is shown in Figure 7B.1. It consists of two poles: positive and negative with respect to ground. Each pole consists of two thyristor converters supplied through a Y-Y and a Y-delta arrangement of transformers to introduce a 30-degree phase shift between the two voltage sets, as discussed in Chapter 6. In this current-link system, the transmission-line inductance on the DC side is usually supplemented by some extra inductance in series by the smoothing reactor, as shown in Figure 7B.1. The current in the DC link cannot change instantaneously because of these inductances: hence the name *current link*. Each pole at the sending end and the receiving end consists of converters consisting of thyristors, which are sometimes referred to by their trade name as *silicone-controlled rectifiers* (SCRs). The characteristic of these converters is explored further in the following subsection.

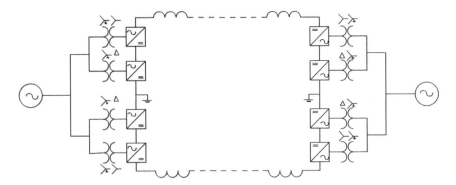

FIGURE 7B.1   Block diagram of a current-link HVDC system.

7B.1   THYRISTOR CONVERTERS

We are familiar with diodes that block a negative polarity voltage so that no current can flow in the reverse direction. Diodes begin to conduct current in the forward direction when a forward polarity voltage appears across them, with only a small voltage drop (on the order of a volt or two) across them. These are represented by a symbol, as shown in Figure 7B.2a, by an anode (A), a cathode (K), and a gate terminal (G). Unlike diodes, thyristors are four-layer devices, as shown in Figure 7B.2b.

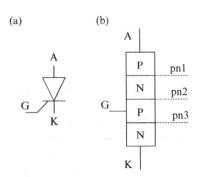

FIGURE 7B.2   (a) Thyristor; (b) layers inside the thyristor.

Similar to diodes, thyristors conduct current only in the forward direction and block negative polarity voltage; but unlike diodes, they can also block a forward-polarity voltage from conducting current, as illustrated next.

Consider a primitive circuit to convert AC into DC, as shown in Figure 7B.3a, with a thyristor in series with an *R–L* load.

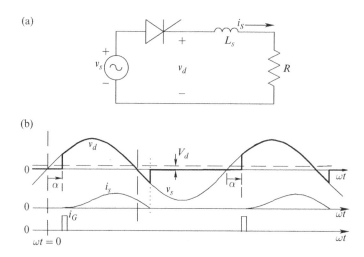

FIGURE 7B.3 (a) Thyristor circuit with an RL load; (b) voltage and current waveforms.

As shown in Figure 7B.3b, beginning at $\omega t = 0$ during the positive half-cycle of the input voltage, a forward voltage appears across the thyristor (anode *A* is positive with respect to cathode *K*); if the thyristor were a diode, a current would begin to flow in this circuit at $\omega t = 0$. This instant at which current would begin to flow if it were a diode is referred to as the *instant of natural conduction*. With the thyristor blocking the forward voltage, the start of conduction can be controlled (delayed) with respect to $\omega t = 0$ by a delay angle α at which instant a current pulse to the gate of the thyristor is applied. Once in the conducting state, the thyristor latches on and behaves like a diode with a very small voltage drop, on the order of 1 to 2 volts, across it (we will idealize it as zero), and the *R–L* load voltage v_d equals v_s, as shown in Figure 7B.3b.

The current waveform in Figure 7B.3b shows that due to the inductor in series, the current through the thyristor keeps flowing for an interval into the negative half-cycle of the input voltage, coming to zero and staying at zero; it cannot reverse in direction due to the thyristor property during the remainder of this half-cycle. In the next voltage cycle, the current conduction again depends on the instant during the positive half-cycle at which the gate pulse is applied. By controlling the delay angle (or *phase control*, as it is often called), we can control the average voltage v_d across the *R–L* load. This principle can be extended to the practical circuits discussed next.

In HVDC systems, each converter uses six thyristors, as shown in Figure 7B.4a. To explain the principle of converter operation, we initially assume that the converter is supplied by ideal three-phase voltage sources, as shown in Figure 7B.4b; the thyristors are drawn in a top group and a bottom group, and the DC side is represented by a DC current source I_d.

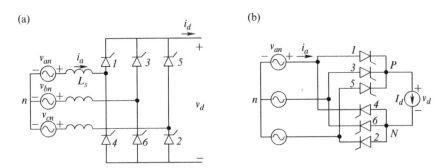

FIGURE 7B.4   Three-phase full-bridge thyristor converter. (a) Actual circuit; (b) idealized circuit.

Initially, we will assume that the thyristors in Figure 7B.4b are replaced by diodes. Diodes represent thyristor operation with the delay angle $\alpha = 0$ degrees; therefore, by replacing thyristors with diodes, we will be able to calculate the DC-side voltage for $\alpha = 0$ degrees. Assuming diodes in Figure 7B.4b, in the top group, all the diodes have their cathodes connected; therefore only the diode with its anode connected to the highest phase voltage conducts, and the other two become reverse-biased. In the bottom group, all the diodes have their anodes connected, and therefore only the diode with its cathode connected to the lowest voltage conducts, and the other two are reverse-biased. The waveforms at points P and N, with respect to the source neutral, are shown in Figure 7B.5a, where the DC-side voltage

$$v_d = v_{Pn} - v_{Nn} \tag{7B.1}$$

is a line-line voltage within each 60-degree interval and is plotted in Figure 7B.5b. The average DC-side voltage with $\alpha = 0$ can be calculated from the waveforms in Figure 7B.5b by considering a 60-degree ($\pi/3$ rad) interval at which the waveforms repeat, where $\sqrt{2}\,V_{LL}$ is the peak value of the line-line input voltage as shown in Figure 7B.5b:

$$V_{do} = \frac{1}{\pi/3} \int_{-\pi/6}^{\pi/6} \sqrt{2}V_{LL} \cos \omega t \cdot d(\omega t) = \frac{3\sqrt{2}}{\pi} V_{LL} \tag{7B-2}$$

This average voltage is indicated in Figure 7B.5b by a straight line as V_{do}, where the subscript o refers to $\alpha = 0$. The line current waveforms are shown in Figure 7B.5c: for example, $i_a = I_d$ during the interval when phase a has the highest voltage and diode 1 connected to it is conducting. Similarly, $i_a = -I_d$ during the interval when phase a has the lowest voltage and diode 4 connected to it is conducting.

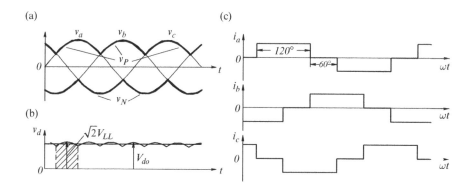

FIGURE 7B.5   Waveforms in a three-phase rectifier with $L_s = 0$ and $\alpha = 0$. (a) Voltage waveforms; (b) output voltage waveform; (c) line current waveforms.

Delaying the gate pulses to the thyristors by an angle α measured with respect to their instants of natural conductions (the instant of natural conduction for a thyristor is the instant at which the current through it would begin to flow if α was zero), the waveforms are shown in Figure 7B.6.

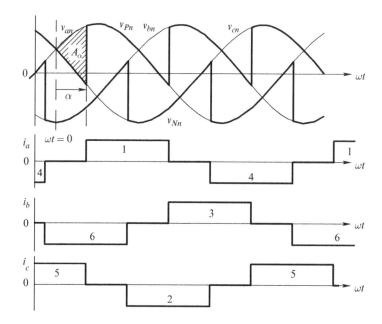

FIGURE 7B.6   Waveforms with $L_s = 0$.

In the DC-side output voltage waveforms, the area A_α corresponds to loss in units of volt-radians due to delaying the gate pulses by α every $\pi/3$ radian. Assuming the time origin in Figure 7B.6 is the instant at which the phase voltage waveforms for phases a and c cross, the line-line voltage v_{ac} waveform can be expressed as $\sqrt{2}V_{LL}\sin\omega t$. Therefore, from Figure 7B.6, the drop ΔV_α in the average DC-side voltage can be calculated as

$$\Delta V_\alpha = \frac{1}{\pi/3}\underbrace{\int_0^\alpha \sqrt{2}V_{LL}\sin\omega t \cdot d(\omega t)}_{A_\alpha} = \frac{3\sqrt{2}}{\pi}V_{LL}(1-\cos\alpha) \qquad (7B.3)$$

Therefore, using Equations 7B.2 and 7B.3, the average value of the DC-side voltage can be controlled by the delay angle as

$$V_{d\alpha} = V_{do} - \Delta V_\alpha = \frac{3\sqrt{2}}{\pi}V_{LL}\cos\alpha \qquad (7B.4)$$

which, as shown by Equation 7.4, is positive for α between 0 and 90 degrees. Hence the converter is said to operate as a *rectifier*, whereas for α beyond 90 degrees, $V_{d\alpha}$ becomes negative, and the converter operates as an *inverter*.

Example 7B.1
The three-phase thyristor converter in Figure 7B.2b is operating in its inverter mode with α = 150 degrees. Draw waveforms similar to Figure 7B.6 for this operating condition.

Solution These waveforms for α = 150 degrees in the inverter mode are shown in Figure 7B.7. On the DC side, the waveform of the voltage v_{Pn} is negative, and that of v_{Nn} is positive. Therefore, the DC-side voltage $v_{PN}(= v_{Pn} - v_{Nn})$ and its average values are negative in the inverter mode of operation. Since the DC-side current is in the same direction, the flow of power is from the DC side to the AC side. On the AC side, phase-current waveforms are shifted (lagging) by α = 150 degrees compared to the waveforms corresponding to $\alpha = 0$.

As shown in Figure 7B.8a, for α = 90 degrees, the converter operates as a rectifier, and the power flows from the AC side to the DC side, as shown in Figure 7B.8b. The opposite is true in the inverter mode with α = 90 degrees. In the inverter mode, the delay angle is limited to approximately 160 degrees, as shown in Figure 7B.8a, due to the commutation angle required to safely commutate current from one thyristor to the next, as discussed next.

In the idealized case with $L_s = 0$, the AC-side currents commutate instantly from one thyristor to another, as shown by the current waveforms in Figure 7B.6. However, in the presence of the AC-side inductance L_s, shown in Figure 7B.4a, it takes a finite interval u during which the current commutates from one thyristor to another, as shown in Figure 7B.9. During this commutation

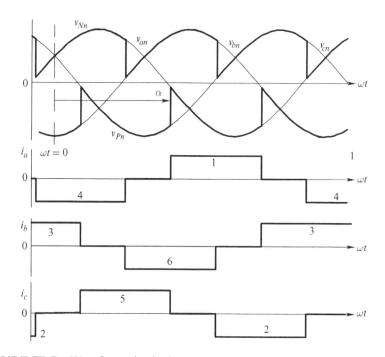

FIGURE 7B.7 Waveforms in the inverter mode.

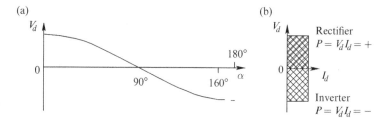

FIGURE 7B.8 (a) Average DC-side voltage as a function of α; (b) converter operation region.

interval u, from α to $\alpha + u$, the instantaneous DC voltage is reduced due to the voltage drop v_L across the inductance in series with the thyristor to which the current is commutating from 0 to $(+I_d)$. Therefore, the average DC output voltage is reduced by an additional area A_u every $\pi / 3$ radians, as shown in Figure 7B.9, where

$$A_u = \int_{\alpha}^{\alpha + u} v_L \, d(\omega t) = \omega L_s \int_0^{I_d} di_s = \omega L_s I_d$$

$$(7B.5)$$

and therefore an additional voltage drop due to the presence of L_s is

$$\Delta V_u = \frac{A_u}{\pi/3} = \frac{3}{\pi}\omega L_s I_d \qquad (7B.6)$$

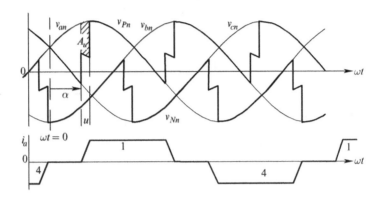

FIGURE 7B.9 Waveforms with L_s.

Therefore, the DC-side output voltage can be written as

$$V_d = V_{d\alpha} - \Delta V_u \qquad (7B.7)$$

Substituting the results from Equations 7B.4 and 7B.6 into Equation 7B.7,

$$V_d = \frac{3\sqrt{2}}{\pi}V_{LL}\cos\alpha - \frac{3}{\pi}\omega L_s I_d \qquad (7B.8)$$

In Figure 7B.9, v_{Pn} during the commutation interval u is the average of v_{an} and v_{cn}.

   Therefore, Equation 7B.8 can also be written as

$$V_d = \frac{3\sqrt{2}}{\pi}V_{LL}\cos(\alpha+u) + \frac{3}{\pi}\omega L_s I_d \qquad (7B.9)$$

Considering the AC side of the converter, in Figure 7B.9, the current (for example, in phase a) can be approximated as a trapezoid staring at an angle $(\pi/6+\alpha)$ rads and rising linearly from $-I_d$ to $+I_d$ during the commutation angle u. With this approximation of a trapezoidal waveform, the fundamental component i_{a1} of the phase current lags behind the phase voltage by an angle $\phi_1 (\approx \alpha + u/2)$, as shown in Figure 7B.10a in the rectifier mode and Figure 7B.10b in the inverter mode. The power factor, always lagging, is as follows:

$$\text{power factor } (PF) \simeq \cos(\alpha + u/2) \qquad (7B.10)$$

The three-phase reactive power consumed by the converter is

$$Q_{3\phi} \simeq 3V_a I_{a1} \sin(\alpha + u/2) \qquad (7B.11)$$

In Equation 7B.11, for approximate calculations, the AC-side current waveforms can be assumed to be rectangular (that is, $u = 0$), in which case $\hat{I}_{a1} = \dfrac{\sqrt{12}}{\pi} I_d$.

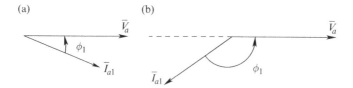

(a)   (b)

FIGURE 7B.10 Power-factor angle. (a) Current and voltage in rectifier mode; (b) current and voltage in inverter mode.

Avoiding Commutation Failure: The commutation of current from one thyristor to the new incoming thyristor is facilitated by the line-line AC voltage corresponding to them, called the *commutating voltage*. In the inverter mode, if α is too large, then $(\alpha + u)$ may exceed 180 degrees. However, beyond 180 degrees, the polarity of the commutating voltage reverses, and the current commutation is not successful. Therefore, to avoid this commutation failure, conservatively, α is limited to 160 degrees or so.

In the waveforms in Figure 7B.6, there are six pulses, each with an interval of 60 degrees ($\pi / 3$ rad) every line-frequency cycle, and hence the converters for which these waveforms are drawn are called *six-pulse converters*. In most HVDC converters, in each pole, a 30-degree phase shift is introduced by the Y-Y and Y-delta transformer arrangements, as shown in Figure 7B.1 and the one-line diagram in Figure 7B.11 of an actual HVDC system.

Each converter draws currents with a six-pulse waveform, as shown in Figure 7B.12a by $i_a(Y - Y)$ and $i_a(Y - \Delta)$. But the sum of these two currents, i_a, drawn from the utility, has a 12-pulse waveform with much lower ripple content than the 6-pulse waveform. Similarly, as shown in Figure 7B.12b, the DC-side voltage waveforms of the two converters, v_{d1} and v_{d2}, in a pole sum up to a 12-pulse waveform v_d that has far fewer ripples on the DC voltage than the 6-pulse waveform. Despite this reduction in ripple (or the *harmonics*, as they are called, of the fundamental line frequency), filters are placed on the AC side for the harmonics in currents and on the DC side for the harmonics in the DC-side voltage, as shown in

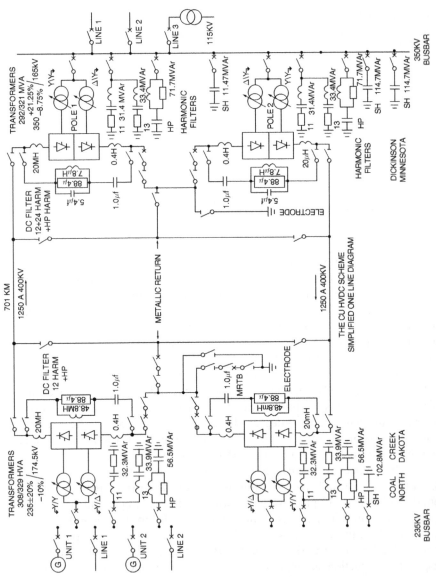

FIGURE 7B.11 HVDC CU project [7B1].

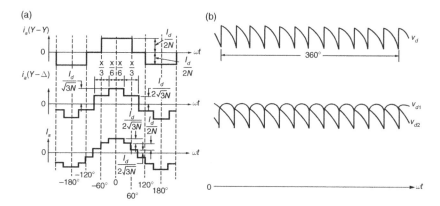

FIGURE 7B.12 Six-pulse and 12-pulse current and voltage waveforms [7B2]. (a) Current waveforms; (b) voltage waveforms.

the one-line diagram in Figure 7B.11 of an actual project, so that these harmonic currents and voltages do not interfere with the power system operation.

REFERENCES

7B1. Great River Energy. www.greatriverenergy.com.

7B2. E. W. Kimbark. 1971. *Direct Current Transmission*, Vol. 1. Wiley Interscience.

PROBLEMS

7.1   In a three-phase thyristor converter, $V_{LL} = 460\,$V(rms) at 60 Hz, and $L_s = 5\,$mH. The delay angle $\alpha = 30$ degrees. This converter supplies 5 kW of power. The DC-side current i_d can be assumed to be purely DC. (a) Calculate the commutation angle u. (b) Draw the waveforms for the converter variables: phase voltages, phase currents v_{Pn}, v_{Nn}, and v_d. (c) Assuming that the currents through the thyristors increase/decrease linearly during commutations, calculate the reactive power drawn by the converter.

7.2   In Figure 7.8a, assume that $L_s = 0$ and $V_{LL} = 480\,$V (rms) at a frequency of 60 Hz. The delay angle $\alpha = 0$ degrees. It supplies power of 10 kW. Calculate and plot the waveforms similar to those in Figure 7.10.

7.3   Repeat Problem 7.2 if the delay angle $\alpha = 45$ degrees.

7.4   Repeat Problem 7.2 if the delay angle $\alpha = 145$ degrees.

7.5   Repeat Problem 7.2 if the AC-side inductance L_s is such that the commutation angle $u = 10$ degrees.

7.6 Repeat Problem 7.2 if the delay angle α = 45 degrees and the AC-side inductance L_s is such that the commutation angle u = 10 degrees.

7.7 Repeat Problem 7.2 if the delay angle α = 145 degrees and the AC-side inductance L_s is such that the commutation angle u = 10 degrees. Note that the power flow is from the DC side to the AC side.

7.8 Calculate the reactive power consumed by the converter and the power factor angle in Problem 7.5.

7.9 Calculate the reactive power consumed by the converter and the power factor angle in Problem 7.6.

7.10 Calculate the reactive power consumed by the converter and the power factor angle in Problem 7.7. Note that the power flow is from the DC side to the AC side.

7.11 In the block diagram in Figure 7.17, for both three-phase converters, $V_{d0} = 480\,\text{kV}$. The DC-side current is $I_d = 1\,\text{kA}$. Converter 2, operating as an inverter, establishes the DC-link voltage such that $V_{d2} = -425\,\text{kV}$. $R_d = 10.0\,\Omega$. The drop in DC voltage due to commutation overlap in each converter is 10 kV. In DC steady state, calculate the following angles: α_1, α_2, u_1, and u_2.

7.12 In Problem 7.11, calculate the reactive power drawn by each converter. Assume that the currents through the thyristors increase/decrease linearly during commutations.

7.13 In a voltage-link converter, the DC-bus voltage V_d and the three-phase voltages to be synthesized are given. (a) Write the expressions for v_{aN}, v_{bN}, and v_{cN}. (b) Write the expressions for d_a, d_b, and d_c.

7.14 In Figure 7.20a, $\bar{V}_{bus} = 1\angle 0\,\text{pu}$ and $X_L = 0.1\,\text{pu}$. Calculate $\bar{V}_{conv}$ to supply 1 pu power at V_{bus} with (a) $Q = 0.5\,\text{pu}$ and (b) $Q = -0.5\,\text{pu}$.

8

DISTRIBUTION SYSTEM, LOADS, AND POWER QUALITY

8.1 INTRODUCTION

In this chapter, we will briefly examine the distribution system, the nature of prominent types of loads, and power quality considerations to keep the power supply voltage sinusoidal at the rated voltage and line frequency.

8.2 DISTRIBUTION SYSTEMS

Unlike distributed generation (DG) in the future, electricity today is normally generated remotely to the load centers and is transported through transmission lines at high, extra-high, and even ultra-high voltages. Electricity from this transmission network is passed on to the subtransmission network at voltages from 230 kV down to 35 kV. Some utilities consider this subtransmission network part of the distribution system, while others consider only the network below 35 kV to be the distribution system. As stated in Chapter 4, approximately 9% of the electricity in the United States is dissipated in transmission and distribution systems, most of it later. Most industrial, commercial, and residential loads are supplied at much lower voltages than 34.5 kV, with the primary voltages in a range of 34.5 kV (19.92 kV per-phase) down to

Electric Power Systems with Renewables: Simulations Using PSS®E, Second Edition. Ned Mohan and Swaroop Guggilam.
© 2023 John Wiley & Sons, Inc. Published 2023 by John Wiley & Sons, Inc.
Companion Website: www.wiley.com/go/mohaneps

12 kV (6.93 kV per-phase) and the secondary voltages at 480/277 V, three-phase, four-wire, and 120/240 V single-phase. Other secondary voltages are 208/120 V, 480 V, and 600 V [1].

For residential loads, distribution substations supply single-phase distribution lines for various communities at voltages such as 13.8 kV. These voltages are stepped down locally to supply a set of houses at ±120 V, as shown in Figure 8.1. At the entrance of a house, the neutral is grounded. From the central breaker, a set of circuits, each with its own circuit breaker, are supplied by the +120 V line conductor and the neutral, whereas the other set is supplied by the –120 V line conductor and the neutral.

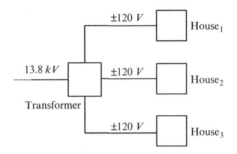

FIGURE 8.1   Residential distribution system.

The ground conductor is carried along with each circuit to three-prong outlets: one for the line conductor, the second for the neutral, and the third for the ground conductor that is connected to the chassis of the load, such as a toaster. The reason for carrying along the ground conductor is the elimination of shock hazards. Normally there is no current flow through the ground conductor, and it remains at ground potential, whereas the neutral conductor can rise above the ground potential due to current flow through its impedance. (Note that only a current of 5 mA is needed through the human heart to make it go into ventricular defibrillation.) Outlets in more shock-susceptible areas with wet surfaces are equipped with ground fault interrupters (GFIs) that measure the difference between the current on the line conductor and that returning on the neutral conductor. The difference between the two currents indicates a fault to ground, and the GFI triggers the associated circuit breaker to be tripped.

8.3   POWER SYSTEM LOADS

Power systems are designed to serve industrial, commercial, and residential loads. A plot of the power demand on a utility, as a function of the time of day, is plotted in Figure 8.2a as an example. The waveform of this load curve may be different for a weekday compared to weekends, reflecting the shutdown of factories and commercial shops. The area underneath the plot in Figure 8.2a represents energy that the utility must supply in 24 hours, whereas the peak of this waveform is the peak load that the utility must supply by its own generation or by purchasing power from other utilities. The ratio between the kilowatt-hours represented under the load curve in Figure 8.2a and the kilowatt-hours that would be necessary to generate if the load were constant at its peak value over the entire 24-hour period is called the *load factor*. The annual load-duration curve in Figure 8.2b [2] shows that the load is at or 90% above its peak value for only a small percentage of the time in a year.

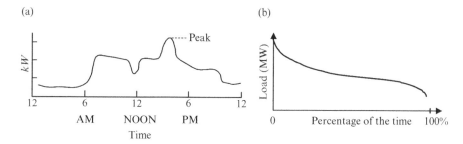

(a)   (b)

FIGURE 8.2 (a) Daily load vs. time; (b) load duration curve. Source: [2].

Ideally, utilities would like the load factor to be unity, but in reality, it is far less. Utilities desire a load factor of unity because peak power is expensive to generate or purchase from other utilities. It taxes the transmission system capacity and results in extra power losses. Some utilities have successfully increased the load factor on their systems by incentivizing customers to shift their loads to off-peak periods by offering lower electricity rates at off-peak periods. Others have used energy storage in the form of pumped hydro: for example, during off-peak periods, water is pumped from a lower reservoir up to a higher reservoir to generate electricity during peak periods.

Utilities also invest considerable sums of money in load forecasting over the next 24 hours to make purchasing agreements and commit units to meet the expected load demand. Accurate forecasting of load results in considerable savings.

8.3.1 Nature of Power System Loads

Figure 8.3a shows the percentage of electricity consumed by the industrial, commercial, and residential sectors in the United States, and Figure 8.3b shows electricity consumed by various types of loads.

Most utilities serve a variety of such loads. Industrial loads mainly depend on the type of industry, whereas commercial and residential loads generally consist of the following in a variety of mixes:

- Electrical heating
- Lighting (incandescent and fluorescent)

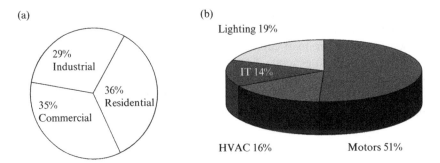

FIGURE 8.3 (a) Percentage of electricity consumed in various sectors in the United States; (b) percentage of electricity consumed by various types of load in the United States.

- Motor loads to drive compressors for heating, ventilating, and air conditioning (HVAC)
- Power electronics-based compressor loads and compact fluorescent lighting (CFL)

Each of these loads behaves differently, given changes in the voltage magnitude and frequency. Generally, frequency does not vary appreciably in a large interconnected system such as that in North America (unless a phenomenon called *islanding* occurs) to be concerned. However, the voltage sensitivity of loads must be included when calculating power flow and stability. Utilities estimate the mix of loads on their systems. Knowing how each load behaves, the load aggregate is modeled in various studies by a combination of constant impedance, constant power, and constant current representations.

Voltage sensitivities $a(=\partial P/\partial V)$ to real power and $b(=\partial Q/\partial V)$ to reactive power of various loads can be approximated as follows and are summarized in Table 8.1:

- *Electric heating*: These loads are resistive. Therefore their power factor equals unity; $a = 2$ and $b = 0$.
- *Incandescent lighting*: Being resistive, their power factor is unity. Because the filament resistance is nonlinear, $a = 1.5$ and $b = 0$.
- *Fluorescent lighting*: These use magnetic ballasts, and their power factor is approximated at 0.9. It is reported that for such loads, $a = 1$ and $b = 1$.
- *Motor loads*: Single-phase motors are used in smaller power ratings and three-phase motors in larger power ratings. Their power factor can be approximated in a range of 0.8 to 0.9. Their sensitivities to voltages depend on the type of load being driven, the fan type or compressor type, because of the variation of the torque required by the load as a function of speed. Motor sensitivities are reported to be in a range of $a = 0.05 - 0.5$ and $b = 1 - 3$.

- *Modern power-electronics-based loads*: Modern and future power-electronics-based loads that are increasingly being used are nonlinear; they continue to draw the same power even if the input voltage changes slightly in its magnitude and frequency, as discussed in this section. They can be designed with a power-factor-corrected (PFC) interface that results in essentially a unity power factor. In such loads, ideally, it is possible to get $a = 0$ and $b = 0$. The structure of these loads is described in further detail in the next section.

TABLE 8.1 Approximate power factor and voltage sensitivity of various loads

Type of load	Power factor	$a = \partial P / \partial V$	$b = \partial Q / \partial V$
Electric heating	1.0	2.0	0
Incandescent lighting	1.0	1.5	0
Fluorescent lighting	0.9	1.0	1.0
Motor loads	0.8–0.9	0.05–0.5	1.0–3.0
Modern power-electronics-based loads	1.0	0	0

8.3.1.1 Power-Electronics-Based Loads

The trend in power system loads is that they are increasingly being supplied through power-electronics interfaces. Doing so increases the overall system efficiency, in some cases as much as 30% – for example, in heat pump systems [3] – and hence has large energy conservation potential. In most cases, these loads are supplied by a voltage-link system, discussed in connection with HVDC transmission systems, as shown in Figure 8.4.

FIGURE 8.4 Voltage-link system for modern and future power-electronics-based loads.

A great deal of lighting load is shifting to compact fluorescent lamps (CFLs), where a power-electronics interface shown in Figure 8.4 is needed to produce high-frequency AC in the range of 30 to 40 kHz at which these lamps operate most efficiently. CFLs are approximately four times more efficient than incandescent lamps. That is, to provide the same illumination, they consume only one-fourth as much electricity, amounting to huge savings. Despite their high initial cost, they are being used in increasing numbers, even in developing countries.

Figure 8.5 shows the per-phase steady-state equivalent circuit of a three-phase induction motor. Conventionally, the speed ω_m of such motors has been controlled by reducing the magnitude V_a of the applied voltage without changing its frequency: that is, by keeping the synchronous speed ω_{syn} unchanged. Operating at large values of slip speed ω_{slip}, which equals $(\omega_{syn} - \omega_m)$, causes large power losses in the rotor circuit, resulting in very low energy efficiency during operation.

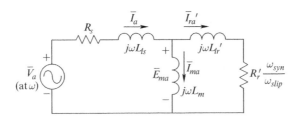

FIGURE 8.5 Per-phase, steady-state equivalent circuit of a three-phase induction motor.

The speed of induction-motor loads can be adjusted efficiently with the power-electronics interface in Figure 8.4, which produces three-phase output voltages whose amplitude and frequency can be controlled independently. By controlling the frequency of the voltages being applied to an induction motor, provided the voltage magnitude is also controlled to result in the rated air-gap flux within the motor, the motor torque-speed characteristics can be plotted, as shown in Figure 8.6.

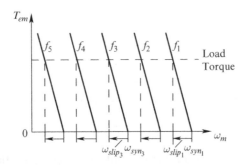

FIGURE 8.6 Torque-speed characteristic of an induction motor at various applied frequencies.

Each frequency of applied voltage results in the corresponding synchronous speed ω_{syn} from which the slip speed ω_{slip} is measured. Assuming a constant-torque load characteristic, as shown by the dotted line in Figure 8.6, the operating speed can be varied continuously by continuously varying the applied frequency. At each operating point, the slip speed, measured with respect to its corresponding synchronous speed, remains small, and hence the energy efficiency remains high.

In the applications mentioned earlier where a regulated DC power supply is needed, the voltage-link system in Figure 8.4 is used, as shown by the block diagram in Figure 8.7, to produce high-frequency AC, which is stepped down in voltage by a high-frequency transformer and then rectified to produce regulated DC output. The transformer operates at high frequency in the range 200 to 300 kHz and hence can be made much smaller than a line-frequency transformer with a similar VA rating. The efficiency of such switch-mode DC power supplies approaches 90%, almost double that associated with linear power supplies.

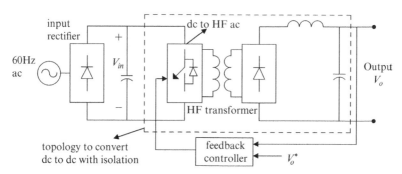

FIGURE 8.7 Switch-mode DC power supply.

In developed countries with a high cost of electricity, most compressor loads, such as air conditioners and heat pumps, are supplied through the power-electronics interface shown in Figure 8.4. Studies have shown that by adjusting the compressor speed to match the thermal load, as much as 30% savings in electricity can be achieved compared to the conventional on/off cycling approach. Similarly, in pump-driven systems to adjust flow rate, controlling the pump speed by using the power-electronics interface in Figure 8.4, considerable improvement in the overall system efficiency can be obtained compared to using a throttling valve. A report by the U.S. Department of Energy (DOE) [4] estimates that if all pump-driven systems were controlled using power electronics, a mature technology now, enough energy could be saved annually to equal that used by the state of New York!

Power-electronics-based loads are often nonlinear in that they often continue to draw the same power irrespective of a change in the input voltage in a small range. If these power-electronics-based loads are not designed appropriately, they draw distorted (non-sinusoidal) currents from the utility source and hence can degrade the quality of power, as discussed in the next section. However, it is possible to design the power-electronics interface in Figure 8.4 with current shaping circuits, often called power-factor-correction (PFC) circuits [5], so that the current drawn from the utility is sinusoidal and at a unity power factor.

8.4 POWER QUALITY CONSIDERATIONS

It is important to customers that the power they receive from their utility is of acceptable quality. These power quality considerations can be classified into the following categories:

- Continuity of service
- Magnitude of voltage
- Voltage waveform

In an interconnected system such as that in North America, the frequency of voltage supply is seldom of concern and therefore is not mentioned here. We will discuss each of the listed considerations in the following subsections.

8.4.1 Continuity of Service

The most serious power quality issue is the lack of continuity of service. Utilities do their best to ensure it because a service disruption also means lost revenue. Interconnected systems, as one of their benefits, improve the continuity of service. If a part of the power system goes down for whatever reason, an interconnected system has a better chance of supplying power through an alternative route.

8.4.1.1 Uninterruptible Power Supplies (UPSs)
To further improve this continuity for critical loads such as some computers and medical equipment, uninterruptible power supplies (UPSs) are used that store energy in chemical batteries and flywheels in the form of inertial energy within the voltage-link system, as shown in Figure 8.8.

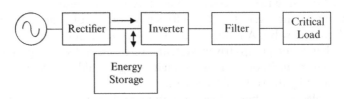

FIGURE 8.8 Uninterruptible power supply.

8.4.1.2 Solid-State Transfer Switches

On a larger scale for an entire industrial plant, it is possible to use two feeders if they are derived from different systems, as shown in Figure 8.9. One of the feeders is used as the primary feeder that normally supplies power to the load. In case of a fault on this feeder, solid-state transfer switches made up of thyristors or IGBTs switch the load to the alternate feeder within a few milliseconds and hence maintain the continuity of service. This action assumes that the same fault, in all probability far away, does not affect both feeders.

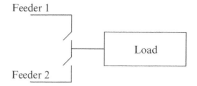

FIGURE 8.9   Alternate feeder.

8.4.2   Voltage Magnitude

Power system loads prefer that the voltage magnitude is at its specified nominal value. Utilities try to maintain voltage supply within a range such as ±5% of the nominal voltage. However, fault conditions (even far away) or heavy or very light loading of transmission lines can cause voltages to be outside the normal range in one or more phases. To maintain this magnitude of voltage, equipment such as a dynamic voltage restorer (DVR) is used. As shown in the block diagram in Figure 8.10, utilizing a voltage-link system, DVRs inject a voltage in series with the utility supply to bring the voltage at the consumer end to within the acceptable range.

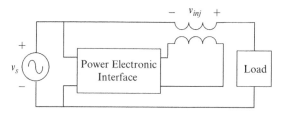

FIGURE 8.10   Dynamic voltage restorer (DVR).

On a larger scale, in distribution substations, voltage regulators and transformers with load tap changing (LTC) automatically try to regulate the supply voltage. According to [6], both three-phase and single-phase voltage regulators are used in distribution substations to regulate the load-side voltage. Substation regulators are one of the primary means, along with load-tap-changing power transformers, shunt capacitors, and distribution line regulators, for maintaining a proper voltage level at a customer's service entrance. A very important function of substation voltage regulation is to correct for supply voltage variation. With the proper use of the control settings and line-drop compensation, regulators can also correct for load variations. A properly applied and controlled voltage regulator not only keeps the voltage at a customer's service entrance within approved limits but also minimizes the range of voltage swings between light and heavy load periods [6].

This problem of voltage outside the normal range can be mitigated by reactive power control using a power-electronics-based device such as a static synchronous compensator (STATCOM), shown in Figure 8.11 as a block diagram. A STATCOM acts as a continually adjustable reactor, which can draw either inductive or capacitive volt-amperes (vars) within its designed rating to regulate the bus voltage. STATCOMs are further explained in Chapter 10.

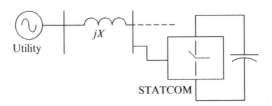

FIGURE 8.11 STATCOM [5].

8.4.3 Voltage Waveform

Most of the load equipment in use is designed assuming a sinusoidal voltage waveform. Any distortion in this voltage waveform can cause loads such as induction motors to draw distorted (non-sinusoidal) currents, which can result in loss of efficiency and overheating and may cause some loads to fail. As mentioned earlier, power-electronics-based loads, if not designed with this consideration, would draw distorted currents and cause the supply voltage, due to its internal impedance, to become distorted and hence cause problems to other adjacent loads.

8.4.3.1 Distortion and Power Factor

To quantify distortion in the current drawn by power-electronics systems, it is necessary to define certain indices.

As a base case, consider the linear $R - L$ load shown in Figure 8.12a, which is supplied by a sinusoidal source in a steady state. The voltage and current phasors are shown in Figure 8.12b, where ϕ is the angle by which the current lags the voltage.

Using rms values for the voltage and current magnitudes, the average power supplied by the source is

$$P = V_s I_s \cos\phi \qquad (8.1)$$

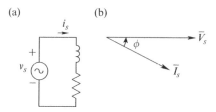

FIGURE 8.12 (a) $R-L$ load with voltage source; (b) phasor diagram.

The power factor PF at which power is drawn is defined as the ratio of the real average power P to the product of the rms voltage and the rms current

$$PF = \frac{P}{V_s I_s} = \cos\phi \quad \text{(using Equation 8.1)} \quad (8.2)$$

where $V_s I_s$ is the apparent power. For a given voltage, from Equation 8.2, the rms current drawn is

$$I_s = \frac{P}{V_s \cdot PF} \quad (8.3)$$

This shows that the power factor PF and the current I_s are inversely proportional. The current flows through the utility distribution lines, transformers, and so on, causing losses in their resistances. This is why utilities prefer unity power factor loads that draw power at the minimum value of the rms current.

8.4.3.2 RMS Value of Distorted Current and Total Harmonic Distortion (THD)
The sinusoidal current drawn by the linear load in Figure 8.12 has zero distortion. However, power-electronics systems without a PFC front-end draw currents with a distorted waveform such as that shown by $i_s(t)$ in Figure 8.13a. The utility voltage $v_s(t)$ is assumed sinusoidal. The following general analysis applies to a utility supply that is either single-phase or three-phase, in which case the analysis is on a per-phase basis.

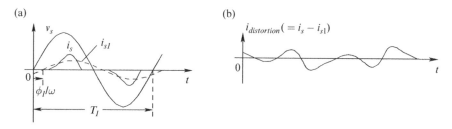

FIGURE 8.13 Current drawn by power-electronics equipment without PFC. (a) Voltage and current waveform; (b) distortion component in the input current.

The current waveform $i_s(t)$ in Figure 8.13a repeats with a time period T_1. By Fourier analysis of this repetitive waveform, we can compute its fundamental frequency $(= 1/T_1)$ component $i_{s1}(t)$, shown as a dotted line in Figure 8.13a. The distortion component $i_{\text{distortion}}(t)$ in the input current is the difference between $i_s(t)$ and the fundamental frequency component $i_{s1}(t)$

$$i_{\text{distortion}}(t) = i_s(t) - i_{s1}(t) \tag{8.4}$$

where $i_{\text{distortion}}(t)$ using Equation 8.4 is plotted in Figure 8.13b. This distortion component consists of components at frequencies that are multiples of the fundamental frequency.

To obtain the rms value of $i_s(t)$ in Figure 8.13a, we apply the basic definition of rms:

$$I_s = \sqrt{\frac{1}{T_1} \int_{T_1} i_s^2(t) \cdot dt} \tag{8.5}$$

Using Equation 8.4,

$$i_s^2(t) = i_{s1}^2(t) + i_{\text{distortion}}^2(t) + 2i_{s1}(t) \times i_{\text{distortion}}(t) \tag{8.6}$$

In a repetitive waveform, the integral of the products of the two harmonic components (including the fundamental) at unequal frequencies over the repetition time period equals zero:

$$\int_{T_1} g_{h_1}(t) \cdot g_{h_2}(t) \cdot dt = 0 \quad h_1 \neq h_2 \tag{8.7}$$

Therefore, substituting Equation 8.6 into Equation 8.5 and making use of Equation 8.7 implies that the integral of the third term on the right side of Equation 8.6 equals zero

$$I_s = \sqrt{\underbrace{\frac{1}{T_1} \int_{T_1} i_{s1}^2(t) \cdot dt}_{I_{s1}^2} + \underbrace{\frac{1}{T_1} \int_{T_1} i_{\text{distortion}}^2(t) \cdot dt}_{I_{s1}^2} + 0} \tag{8.8}$$

or

$$I_s = \sqrt{I_{s1}^2 + I_{\text{distortion}}^2} \tag{8.9}$$

where the rms values of the fundamental frequency component and the distortion component are as follows:

$$I_{s1} = \sqrt{\frac{1}{T_1} \int_{T_1} i_{s1}^2(t) \cdot dt} \qquad (8.10)$$

and

$$I_{\text{distortion}} = \sqrt{\frac{1}{T_1} \int_{T_1} i_{\text{distortion}}^2(t) \cdot dt} \qquad (8.11)$$

Based on the rms values of the fundamental and distortion components in the input current $i_s(t)$, a distortion index called the total harmonic distortion (THD) is defined in percentage as follows:

$$\% \text{THD} = 100 \times \frac{I_{\text{distortion}}}{I_{s1}} \qquad (8.12)$$

Using Equation 8.9 in Equation 8.12,

$$\% \text{THD} = 100 \times \frac{\sqrt{I_s^2 - I_{s1}^2}}{I_{s1}} \qquad (8.13)$$

The rms value of the distortion component can be obtained based on the harmonic components (except the fundamental) as follows, using Equation 8.7,

$$I_{\text{distortion}} = \sqrt{\sum_{h=2}^{\infty} I_{sh}^2} \qquad (8.14)$$

where I_{sh} is the rms value of the harmonic component h.

8.4.3.3 Obtaining Harmonic Components by Fourier Analysis

By Fourier analysis, any distorted (non-sinusoidal) waveform $g(t)$ that is repetitive with a fundamental frequency f_1, for example i_s in Figure 8.13a, can be expressed as a sum of sinusoidal components at the fundamental and its multiple (harmonic) frequencies

$$g(t) = G_0 + \sum_{h=1}^{\infty} g_h(t) = G_0 + \sum_{h=1}^{\infty} \{a_h \cos(h\omega t) + b_h \sin(h\omega t)\} \qquad (8.15)$$

where the average value G_0 is DC:

$$G_0 = \frac{1}{2\pi} \int_0^{2\pi} g(t) \cdot d(\omega t) \qquad (8.16)$$

The sinusoidal waveforms in Equation 8.15 at the fundamental frequency f_1 ($h = 1$) and the harmonic components at frequencies h times f_1 can be expressed as the sum of their cosine and sine components:

$$a_h = \frac{1}{\pi}\int_0^{2\pi} g(t)\cos(h\omega t)d(\omega t) \quad h = 1,2,...,\infty \qquad (8.17)$$

$$b_h = \frac{1}{\pi}\int_0^{2\pi} g(t)\sin(h\omega t)d(\omega t) \quad h = 1,2,...,\infty \qquad (8.18)$$

The cosine and the sine components given by Equations 8.17 and 8.18 can be combined and written as a phasor in terms of its rms value

$$\overline{G}_h = G_h \angle \phi_h \qquad (8.19)$$

where the rms magnitude in terms of the peak values a_h and b_h equals

$$G_h = \frac{\sqrt{a_h^2 + b_h^2}}{2} \qquad (8.20)$$

and the phase ϕ_h can be expressed as

$$\tan\phi_h = \frac{-b_h}{a_h} \qquad (8.21)$$

It can be shown that the rms value of the distorted function $g(t)$ can be expressed in terms of its average and the sinusoidal components as

$$G = \sqrt{G_0^2 + \sum_{h=1}^{\infty} G_h^2} \qquad (8.22)$$

In Fourier analysis, by appropriate selection of the time origin, it is often possible to make the sine or the cosine components in Equation 8.15, thus considerably simplifying the analysis, as illustrated by a simple example.

Example 8.1
A current i_s with a square waveform is shown in Figure 8.14a. Calculate and plot its fundamental frequency component and its distortion component. What is the %THD associated with this waveform?

Solution From Fourier analysis, by choosing the time origin as shown in Figure 8.14a, $i_s(t)$ in Figure 8.14a can be expressed as

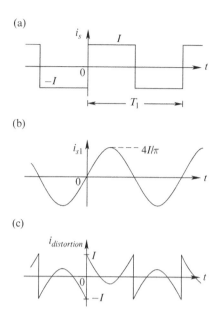

FIGURE 8.14 Current for Example 8.1.

$$i_s = \frac{4}{\pi}I(\sin \omega_1 t + \frac{1}{3}\sin 3\omega_1 t + \frac{1}{5}\sin 5\omega_1 t + \frac{1}{7}\sin 7\omega_1 t + ...) \qquad (8.23)$$

The fundamental frequency component and the distortion component are plotted in Figures 8.14b and 8.14c.

From Figure 8.14a, it is obvious that the rms value I_s of the square waveform is equal to I. In the Fourier expression of Equation 8.23, the rms value of the fundamental frequency component is

$$I_{s1} = \frac{(4/\pi)}{\sqrt{2}}I = 0.9I$$

Therefore, the distortion component can be calculated from Equation 8.9 as

$$I_{distortion} = \sqrt{I_S^2 - I_{S1}^2} = \sqrt{I^2 - (0.9I)^2} = 0.436I$$

Therefore, using the definition of THD,

$$\%\text{THD} = 100 \times \frac{I_{distortion}}{I_{S1}} = 100 \times \frac{0.436I}{0.9I} = 48.4\%$$

8.4.3.4   The Displacement Power Factor (DPF) and Power Factor (PF)

Next, we will consider the power factor at which a load draws power with a distorted current waveform such as that shown in Figure 8.13a. As before, it is reasonable to assume that the utility-supplied line-frequency voltage $v_s(t)$ is sinusoidal, with an rms value of V_s and a frequency $f_1\left(=\dfrac{\omega_1}{2\pi}\right)$. Based on Equation 8.7, which states that the product of the cross-frequency terms has a zero average, the average power P drawn by the load in Figure 8.13a is due only to the fundamental-frequency component of the current:

$$P = \frac{1}{T_1}\int_{T_1} v_s(t)\cdot i_s(t)\cdot dt = \frac{1}{T_1}\int_{T_1} v_s(t)\cdot i_{s1}(t)\cdot dt \tag{8.24}$$

Therefore, in contrast to Equation 8.1 for a linear load, in a load that draws distorted current, similar to Equation 8.1,

$$P = V_s I_{s1}\cos\phi_1 \tag{8.25}$$

where ϕ_1 is the angle by which the fundamental frequency current component $i_{s1}(t)$ lags behind the voltage, as shown in Figure 8.13a.

   At this point, another term called the displacement power factor (DPF) needs to be introduced, where

$$\text{DPF} = \cos\phi_1 \tag{8.26}$$

Therefore, using the DPF in Equation 8.25,

$$P = V_s I_{s1}(DPF) \tag{8.27}$$

In the presence of distortion in the current, the meaning and therefore the definition of the power factor at which the real average power P is drawn remain the same as in Equation 8.2 – that is, the ratio of the real power to the product of the rms voltage and the rms current:

$$PF = \frac{P}{V_s I_s} \tag{8.28}$$

Substituting Equation 8.27 for P into Equation 8.28,

$$PF = \left(\frac{I_{s1}}{I_s}\right)(DPF) \tag{8.29}$$

In linear loads that draw sinusoidal currents, the current ratio (I_{s1}/I_s) in Equation 8.29 is unity, and hence PF = DPF. Equation 8.29 shows that a high distortion in the current waveform leads to a low power factor, even if the DPF is high. Using Equation 8.13, the ratio (I_{s1}/I_s) in Equation 8.29 can be expressed in terms of the total harmonic distortion as

$$\frac{I_{s1}}{I_s} = \frac{1}{\sqrt{1 + \left(\dfrac{\%\text{THD}}{100}\right)^2}} \tag{8.30}$$

Therefore, in Equation 8.29,

$$\text{PF} = \frac{1}{\sqrt{1 + \left(\dfrac{\%\text{THD}}{100}\right)^2}} \cdot \text{DPF} \tag{8.31}$$

The effect of THD on the power factor is shown in Figure 8.15 by plotting (PF / DPF) versus THD. It shows that even if the DPF is unity, a total harmonic distortion of 100% (which is possible in power-electronics systems unless corrective measures are taken) can reduce the power factor to approximately 0.7 (or $\dfrac{1}{\sqrt{2}} = 0.707$, to be exact), which is unacceptably low.

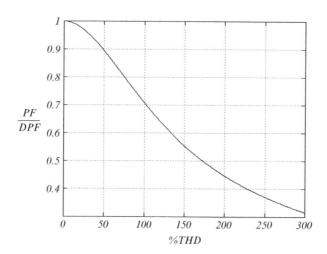

FIGURE 8.15   Relation between PF/DPF and THD.

8.4.3.5   Deleterious Effects of Harmonic Distortion and a Poor Power Factor

There are several deleterious effects of high distortion in the current waveform and the poor power factor that results from it. These are as follows:

- Power loss in utility equipment such as distribution and transmission lines, transformers, and generators increases, possibly to the point of overloading them.
- Harmonic currents can overload the shunt capacitors used by utilities for voltage support and may cause resonance conditions between the capacitive reactance of these capacitors and the inductive reactance of the distribution and transmission lines.
- The utility voltage waveform will also become distorted, adversely affecting other linear loads, if a significant portion of the load supplied by the utility draws power by means of distorted currents.

To prevent degradation in power quality, recommended guidelines (in the form of the IEEE 519) have been suggested by the IEEE. These guidelines place the responsibilities of maintaining power quality on consumers and utilities as follows: (i) on power consumers, such as users of power-electronics systems, to limit the distortion in the current drawn; and (ii) on utilities to ensure that the voltage supply is sinusoidal with less than a specified amount of distortion.

The limits on current distortion placed by the IEEE 519 are shown in Table 8.2, where the limits on harmonic currents, as a ratio of the fundamental component, are specified for various harmonic frequencies. The limits on the THD are also specified. These limits are selected to prevent distortion in the voltage waveform of the utility supply.

TABLE 8.2   Harmonic current distortion (I_h/I_1)

I_{sc}/I_1	Odd harmonic order h (in %)					Total harmonic distortion (%)
	$h < 11$	$11 \leq h < 17$	$17 \leq h < 23$	$23 \leq h < 35$	$35 \leq h$	
<20	4.0	2.0	1.5	0.6	0.3	5.0
20–50	7.0	3.5	2.5	1.0	0.5	8.0
50–100	10.0	4.5	4.0	1.5	0.7	12.0
100–1000	12.0	5.5	5.0	2.0	1.0	15.0
>1000	15.0	7.0	6.0	2.5	1.4	20.0

Therefore, the limits on distortion in Table 8.2 depend on the "stiffness" of the utility supply, which is shown in Figure 8.16a by a voltage source $\bar{V}_s$ in series with internal impedance Z_s. An ideal voltage supply has zero internal impedance. In contrast, the voltage supply at the end of a long distribution line, for example, has a large internal impedance. To define the stiffness of the supply, the short-circuit current I_{sc} is calculated by hypothetically placing short-circuits at the supply terminals, as shown in Figure 8.16b. The stiffness of the supply must be calculated in relation to the load current. Therefore, the stiffness is defined by a ratio called the short-circuit ratio *(SCR)*

$$\text{short-circuit ratio (SCR)} = \frac{I_{sc}}{I_{s1}} \tag{8.32}$$

where I_{s1} is the fundamental frequency component of the load current. Table 8.2 shows that a smaller SCR corresponds to lower limits on the allowed distortion in the current drawn. For a SCR less than 20, the THD in the current must be less than 5%. Power-electronics systems that meet this limit would also meet the limits of stiffer supplies.

FIGURE 8.16 (a) Utility supply; (b) short-circuit current.

Note that IEEE 519 does not propose harmonic guidelines for individual pieces of equipment but rather for the aggregate of loads (such as in an industrial plant) seen from the service entrance, which is also the point of common coupling (PCC) with other customers. However, IEEE 519 is frequently interpreted as the harmonic guidelines for specifying individual pieces of equipment such as motor drives. Other harmonic standards, such as IEC-1000, apply to individual pieces of equipment.

8.4.3.6 Active Filters
To prevent harmonic currents produced by power electronics, loads can be prevented from entering the utility system using filters. These are often passive filters tuned to certain harmonic frequencies: for example, in HVDC terminals. Lately, active filters have also been employed where a current is produced by power electronics and injected into the utility system to nullify the harmonic currents produced by the nonlinear load. Therefore, only a sinusoidal current is drawn from the utility source by combining the nonlinear load and the active filter.

8.5   LOAD MANAGEMENT

This topic is likely to become extremely important in the coming years as the utilities are stretched to meet the load demand. Load management can take many forms [2, 7]. Utilities can implement time-of-day rates, thus giving customers incentives to shift their loads to off-peak hours. They can implement demand-side management (DMS), where certain loads, such as air conditioners, can be interrupted remotely during peak-load hours; in return, customers who sign up for it get rebates on their electricity bills. Large customers can often negotiate to pay a reduced rate for the energy (kWh) used and a demand charge based on the peak power (kW) they draw in a given month. Load shedding based on voltage and frequency can be an important strategy for maintaining proper system operation and preventing voltage collapse and blackouts.

REFERENCES

1. IEEE. Std 141-1986.
2. H. M. Rustebakke (ed). 1983. *Electric Utility Systems and Practices*, 4th ed. John Wiley & Sons.
3. N. Mohan and J. W. Ramsey. 1986. "Comparative Study of Adjustable-Speed Drives for Heat Pumps." EPRI Report.
4. U.S. Department of Energy. www.eia.doe.gov.
5. N. Mohan. 2011. *Power Electronics: A First Course*. John Wiley & Sons. www.wiley.com.
6. U.S. Department of Agriculture, Rural Utilities Service. *Design Guide for Rural Substations*. RUS Bulletin 1724E-300.
7. J. Casazza and F. Delea. 2003. *Understanding Electric Power Systems: An Overview of the Technology and the Marketplace*. IEEE Press and Wiley-Interscience.

PROBLEMS

8.1 Describe the voltage distribution inside homes and residential buildings.

8.2 What is the role of ground fault interrupters (GFIs), and how do they work?

8.3 What is meant by the load factor in describing the daily load curve on utility systems?

8.4 What is the typical make-up of utility loads, and what are their real and reactive power sensitivities to voltages?

8.5 What is the characteristic of power-electronics-based load? Explain.

8.6 A load in a steady state is characterized by $P = 1\text{pu}$ and $Q = 0.5\text{pu}$ at a voltage $V = 1\text{pu}$. Represent it as a constant-impedance load.

8.7 What are the main power quality considerations?

8.8 Describe in words the nature of the CBEMA curve.

8.9 What are uninterruptible power supplies?

8.10 What is meant by a dual-feeder arrangement?

8.11 What are dynamic voltage restorers (DVRs), and how do they work?

8.12 What means are used in substations to regulate voltages?

8.13 What are STATCOMs, and how do they work?

8.14 How is the total harmonic distortion defined in current waveforms?

8.15 What is meant by power-factor-correction (PFC) circuits?

8.16 How do active filters work?

8.17 In a single-phase power-electronics load, $I_s = 10\,\text{A (rms)}$, $I_{s1} = 8\,\text{A (rms)}$, and DPF $= 0.9$. Calculate $I_{\text{distortion}}$, %THD, and PF.

8.18 In a single-phase power electronics load, the following operating condition are given: $V_s = 120\,\text{V(rms)}$, $P = 1\,\text{kW}$, $I_{s1} = 10\,\text{A}$, and THD $= 80\%$. Calculate the following: DPF, $I_{\text{distortion}}$, I_s, and PF.

8.19 What are active filters, and how do they work?

8.20 Define the following terms and their significance: load management, demand-side management, load shedding, demand charges, time-of-day rates, load forecasting, and annual load-duration curve.

9
SYNCHRONOUS GENERATORS

9.1 INTRODUCTION

In most power plants, hydro, steam, or gas turbines provide mechanical input to synchronous generators that convert mechanical input to three-phase electrical power output. Synchronous generators, by means of their voltage excitation system, called the *voltage regulator*, are also the primary means of supporting and regulating the system voltage to its nominal value. Thousands of such generators operate in synchronism in the interconnected system in North America. In this chapter, we will briefly look at the basic structure of synchronous generators and the fundamental principles of the electromagnetic interactions that govern their operation. These synchronous generators can be classified into two broad categories:

- Turbo alternators used with steam turbines, as shown in Figure 9.1a, or with gas turbines that rotate at high speeds, such as at 1,800 rpm for producing 60 Hz output (or 1,500 rpm for 50 Hz output).
- Hydro generators used in hydropower plants, as shown in Figure 9.1b, where turbines rotate at very slow speeds (a few hundred rpm). These generators are very large, with many poles to produce the output at the line frequency of 60 or 50 Hz

We will focus on the high-speed turbo-alternators in Figure 9.1a, although the basic principles involved also apply to the hydro generators. Such generators can be several hundred MVA in ratings at voltages around 20 kV.

Electric Power Systems with Renewables: Simulations Using PSS®E, Second Edition. Ned Mohan and Swaroop Guggilam.

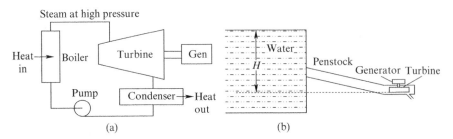

FIGURE 9.1 Synchronous generators driven by (a) steam turbines and (b) hydraulic turbines.

9.2 STRUCTURE

Electrical generators are designed to be long and cylindrical, as shown in Figure 9.2a. To discuss their characteristics, we will describe them based on their cross-section by hypothetically slicing them by a plane perpendicular to the shaft axis in Figure 9.2b. Looking from one side, Figure 9.2b shows the stationary part, called the *stator*, made up of magnetic material such as silicon steel and firmly affixed to the foundation. The rotor on a set of bearings is free to rotate, and the stator and the rotor are separated by a very small air gap.

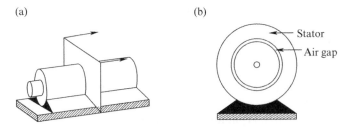

FIGURE 9.2 (a) Three-dimensional representation; (b) cross-section representation.

Windings placed on the stator and rotor produce magnetic flux; Figure 9.3 shows the representation of flux lines in various types of generators. Figures 9.3a and b show a round rotor (non-salient) generator, where the generator is a two-pole generator in Figure 9.3a and a four-pole generator in Figure 9.3b. The cross-section in Figure 9.3c illustrates a four-pole salient pole generator where there are distinct poles, called *salient poles*, on the rotor. Hydro generators may consist of dozens of such pole pairs in a salient construction, where the magnetic reluctance to flux lines is much smaller in the radial path through the rotor-pole axis than in a path between the two adjacent poles.

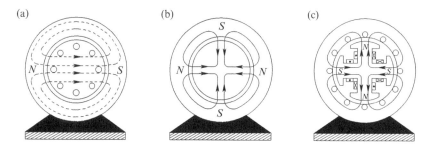

FIGURE 9.3 Machine structure. (a) Two-pole generator; (b) four-pole generator; (c) four-pole salient generator.

In multi-pole generators with more than one pole pair, as shown in Figures 9.3b and 9.3c, it is sufficient to consider only one pole-pair consisting of adjacent north and south poles due to complete symmetry around the periphery in the air gap. Other pole pairs have identical conditions of magnetic fields and currents. Therefore, a multi-pole generator (with the number of poles $p>2$) can be analyzed by considering that pole-pair spans 2π electrical radians and expressing the rotor speed in the units of electrical radians per second to be $p/2$ times its mechanical speed. For ease of explanation in this chapter, we will assume a two-pole generator with $p = 2$.

To minimize the ampere-turns required to create flux lines shown crossing the air gap in Figure 9.3, both the rotor and the stator consist of high-permeability ferromagnetic materials, and the length of the air gap is kept as small as possible, which is shown highly exaggerated for ease of drawing. As in transformers, to reduce eddy-current losses, the stator consists of silicon-steel laminations, which are insulated from each other by a layer of thin varnish. These laminations are stacked together perpendicular to the shaft axis. Conductors, which run parallel to the shaft axis, are placed in the slots cut into these laminations. Hydrogen, oil, or water is often used as the coolant.

9.2.1 Stator with Three-Phase Windings

In synchronous generators, the stator has three-phase windings, with their respective magnetic axes, as shown in Figure 9.4a. Ideally, the windings for each phase should produce a sinusoidally distributed flux density in the air gap in the radial direction. Theoretically, this requires a sinusoidally distributed winding in each phase. In practice, this is approximated in a variety of ways. To visualize this sinusoidal distribution, consider the winding for phase a, shown in Figure 9.4b, where, in the slots, the number of turns per coil for phase a progressively increases away from the magnetic axis, reaching a maximum at $\theta = 90$ degrees. Each coil, such as the coil with sides 1 and 1′, spans 180 degrees, where the current into coil side 1 returns in 1′ through the end turn at the back of the generator. This coil (1,1′) is connected in series to coil side 2 of the next coil (2,2′), and so on. Graphically, these windings are simply drawn, as shown in Figure 9.4a, with the understanding that each is sinusoidally distributed, with its magnetic axis as shown.

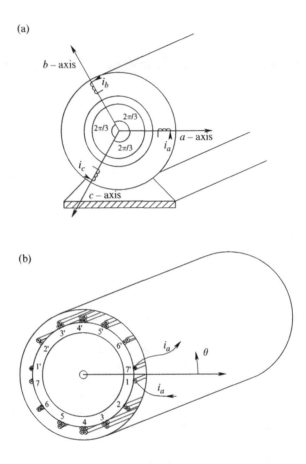

FIGURE 9.4 (a) Synchronous generator with three-phase winding representation; (b) phase *a* only.

In Figure 9.4b, we focused only on phase *a*, which has its magnetic axis along $\theta = 0$ degrees. There are two more identical sinusoidally-distributed windings for phases *b* and c, with magnetic axes along $\theta = 120$ degrees and $\theta = 240$ degrees, respectively, as represented in Figure 9.5a. These three windings are generally connected in a Y arrangement by connecting terminals *a′*, *b′*, and *c′* together, as shown in Figure 9.5b. Flux-density distributions in the air gap due to currents i_b and i_c, identical in shape to that due to i_a, peak along their respective phase *b* and phase *c* magnetic axes.

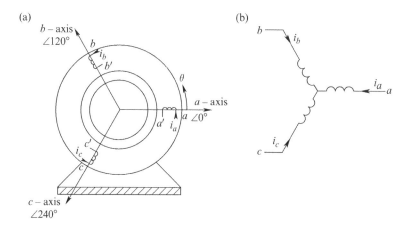

FIGURE 9.5 (a) Cross-section with three phases; (b) connection of three-phase windings.

9.2.2 Rotor with DC Field Winding

The rotor of a synchronous generator contains a field winding in its slots. This winding is supplied by a DC voltage, resulting in a DC current I_f. The field current I_f in Figure 9.6 produces the rotor field in the air gap. It is desired that this field flux density be sinusoidally distributed in the air gap in the radial direction, effectively producing north and south poles, as shown in Figure 9.6. By controlling I_f and hence the rotor-produced field, it is possible to control the induced emf of this generator and the reactive power delivered by it. This field-excitation system is discussed later in this chapter.

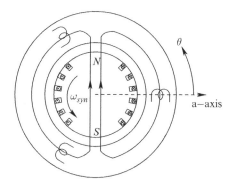

FIGURE 9.6 Field winding on the rotor supplied by a DC current I_f.

In a steady state, the rotor rotates at a speed called the *synchronous speed* ω_{syn} in rad/s, which in a two-pole machine equals $\omega(= 2\pi f)$, where f is the frequency of the generated output voltages.

9.3   INDUCED EMF IN THE STATOR WINDINGS

To discuss induced emf, we will concentrate on phase a, realizing that similar emfs are induced in the other two-phase windings. The phase *a* winding is shown in Figure 9.7 by a single coil, where it is understood that this winding is sinusoidally distributed, as discussed earlier. Current i_a in the direction shown in Figure 9.7 is chosen so that it results in flux lines that peak along the phase *a* magnetic axis. The induced emf polarity in Figure 9.7 is chosen to be positive where the current exits since we are interested in the generator mode of operation (as opposed to motor mode, where following the passive sign convention would require that the current enter the positive terminal of the induced voltage).

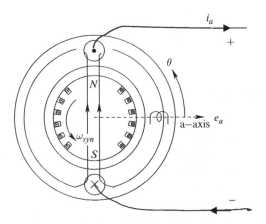

FIGURE 9.7   Current direction and voltage polarity; the rotor position shown induces maximum e_a.

In the stator windings, there are two causes of induced emf, which will be discussed one at a time. Assuming that there is no magnetic saturation, these two induced emfs will be superimposed to yield the resultant induced emf.

9.3.1   Induced EMF Due to Rotation of the Field Flux with the Rotor

In Figure 9.8a, time $t = 0$ is chosen such that the axis of the field winding is vertically up. Due to this field winding, the density of the flux lines in φ_f that links the stator is distributed co-sinusoidally in the radial direction, peaking along the axis of the field winding. This field-flux density distribution can be represented by a space vector $\vec{B}_f$, as shown in Figure 9.8b at time $t = 0$. The length of the space vector represents the value of the flux-density peak, and its angle measured with respect to the phase *a* axis represents its orientation. This space vector is distinguished from phasors by an arrow on top. As the rotor rotates, so does the $\vec{B}_f$ space vector.

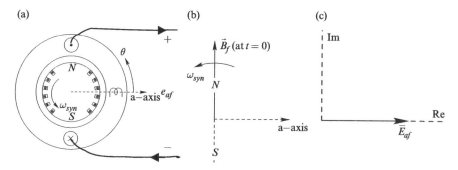

FIGURE 9.8 Induced emf e_{af} due to the rotating rotor field with the rotor. (a) Rotor with field at $t = 0$; (b) space vector diagram of field flux density; (c) phasor diagram of induced emf.

Due to the rotor field-flux lines rotating with the rotor and cutting the stationary windings in the stator, voltages are induced in the stator phase windings in accordance with Faraday's law. At an instant when the rotor and $\vec{B}_f$ orientation are as shown in Figures 9.8a and b, respectively, and rotating at a speed ω_{syn}, the voltage induced is maximum in phase a since the greatest density of field-flux lines is cutting the greatest density of phase a conductors. If the rotor is rotating counterclockwise, then by Lenz's law, we can determine the induced voltage in phase a at $t = 0$ to be positive, with the voltage polarity defined in Figure 9.8a. As the rotor rotates with time, the induced emf in phase a varies co-sinusoidally with time; this voltage is represented by a phasor $\bar{E}_{af}$, as shown in Figure 9.8c.

Both the space vector $\vec{B}_f$ in the space-vector diagram in Figure 9.8b and the phasor $\bar{E}_{af}$ in the phasor diagram in Figure 9.8c are complex variables – that is, they both have amplitudes and angles. Therefore, from Figures 9.8b and c, these two can be related as follows:

$$\bar{E}_{af} = (-j)k_f \vec{B}_f(0) \tag{9.1}$$

where $\vec{B}_f(0)$ is the space vector at time $t = 0$, and k_f is a constant of proportionality depending on the machine construction details and the synchronous speed ω_{syn}. The reason for $(-j)$ in Equation 9.1 is that $\bar{E}_{af}$ lags behind $\vec{B}_f(0)$ by 90 degrees.

9.3.2 Induced EMF Due to the Rotating Magnetic Field (Armature Reaction) Created by the Stator Currents

When a generator is connected to the electrical grid, the result is a flow of sinusoidal phase currents. These currents are necessary to produce power that is supplied to the grid. The flow of these phase currents produces a rotating magnetic field that, in addition to the rotating field flux, also cuts the stator phase windings that are stationary.

In Figure 9.9a, each phase current results in a pulsating flux-density distribution that peaks along its phase axis and is proportional to the instantaneous value of the phase current; at any instant, the flux density drops off co-sinusoidally away from its phase axis. Therefore, the flux-density distribution produced by each phase winding can be represented by a space vector oriented along the respective phase axis. Each of these space vectors is stationary in position, but its amplitude pulsates with time as the phase-current value changes with time and can be expressed as

$$\vec{B}_{i_a} = (k_1 i_a)e^{j0} \quad \vec{B}_{i_b} = (k_1 i_b)e^{j2\pi/3} \quad \vec{B}_{i_c} = (k_1 i_c)e^{j4\pi/3} \tag{9.2}$$

where k_1 is a machine constant that relates the instantaneous phase-current and peak flux-density values. The resultant flux-density distribution can be obtained by vectorially summing the three space vectors, using the principle of superposition by assuming a linear magnetic circuit:

$$\vec{B}_{AR} = k_1(i_a e^{j0} + i_b e^{j2\pi/3} + i_c e^{j4\pi/3}) \tag{9.3}$$

where the subscript AR refers to the *armature reaction*, as it is commonly called.

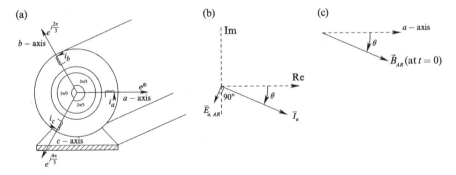

FIGURE 9.9 (a) Armature reaction due to stator currents; (b) phasor diagram; (c) space vector diagram.

Let the three-phase currents be as follows, where $t = 0$ corresponds to the rotor position shown in Figure 9.8a, ω (= $2\pi f$) is the frequency of the stator voltages and currents, and θ is an angle by which the phase current lags the generated voltage $\bar{E}_{af}$ in Figure 9.8c:

$$i_a = I_a \cos(\omega t - \theta) \; i_b = I_a \cos(\omega t - \theta - 2\pi/3) \; i_c = I_a \cos(\omega t - \theta - 4\pi/3) \quad (9.4)$$

where I_a is the RMS value of each phase current. The sinusoidal current i_a is represented by a phasor $\bar{I}_a$ in Figure 9.9b; similarly, other phase currents can be represented as phasors. Note that in a steady state,

$$\omega = \omega_{\text{syn}} \quad (9.5)$$

By substituting the current expressions of Equation 9.4 into Equation 9.3 and making use of Equation 9.5,

$$\vec{B}_{\text{AR}} = (k_2 I_a)e^{j(\omega_{\text{syn}}t - \theta)} \quad (9.6)$$

recalling that $\vec{B}_{\text{AR}}$ is the space vector representing the flux-density distribution due to all three phase currents at any instant t. In a steady state, the peak of this space vector remains constant (k_2 is another constant) for a given current rms magnitude I_a, and it rotates counterclockwise at the synchronous speed ω_{syn} with time. From Equations 9.4 and 9.6, we should note that the orientation of $\vec{B}_{\text{AR}}$ at time $t = 0$ in the space-vector diagram in Figure 9.9c is the same as the orientation of $\bar{I}_a$ in the phasor diagram in Figure 9.9b.

   Just as the rotating field-flux density distribution $\vec{B}_f$ induces field voltages $\bar{E}_{af}$ in the stationary stator phase a winding, rotating $\vec{B}_{\text{AR}}$ induces armature-reaction voltages $\bar{E}_{a,\text{AR}}$; similar voltages are induced in the other two phases, b and c, time-delayed by 120 degrees and 240 degrees, respectively. Therefore, making the analogy with Equation 9.1, $\bar{E}_{a,\text{AR}}$ (Figure 9.9b) lags $\vec{B}_{\text{AR}}(0)$ (Figure 9.9c) by 90 degrees, and it can be expressed as follows:

$$\bar{E}_{a,\text{AR}} = (-j)k_3 \vec{B}_{\text{AR}}(0) \quad (9.7)$$

where k_3 is another constant of proportionality. Figures 9.9b and 9.9c show that the orientation of $\vec{B}_{\text{AR}}(0)$ at $t = 0$ in the space-vector diagram is the same as that of $\bar{I}_a$ in the phasor diagram, and the magnitudes of both these variables are related through Equations 9.6 and 9.7. Therefore, the armature-reaction voltage in phase a can be written as

$$\bar{E}_{a,\text{AR}} = -jX_m \bar{I}_a \quad (9.8)$$

Equation 9.8 shows that $\bar{E}_{a,\text{AR}}$ lags $\bar{I}_a$ by 90 degrees, and the magnitudes of these two are related to each other by what is called the *magnetizing reactance* X_m of the synchronous generator.

9.3.3 Combined Induced EMF Due to the Field Flux and Armature Reaction

We saw in the previous two subsections that the induced emf in phase a (and similarly in phases b and c) is due to two mechanisms. In a magnetic circuit, assuming no saturation, these two emfs can be combined to determine the resultant emf:

$$\bar{E}_a = \bar{E}_{af} + \bar{E}_{a,\mathrm{AR}} = \bar{E}_{af} - jX_m\bar{I}_a \tag{9.9}$$

as shown in Figure 9.10a, where $\bar{I}_a$ lags $\bar{E}_{af}$ by the same angle θ as defined in Equation 9.4. The relationship in Equation 9.9 can be represented by a per-phase equivalent shown in Figure 9.10b. Including the effect of the leakage flux by a voltage drop across the leakage reactance $X_{\ell s}$ and including the voltage drop across the phase-winding resistance R_s, we can write the terminal voltage expression as

$$\bar{V} = \bar{E}_{af} - jX_s\bar{I}_a - R_s\bar{I}_a \tag{9.10}$$

where $X_s(= X_{\ell s} + X_m)$ is called the *synchronous reactance*, which is the sum of the leakage reactance $X_{\ell s}$ of each stator winding and the magnetizing reactance X_m.

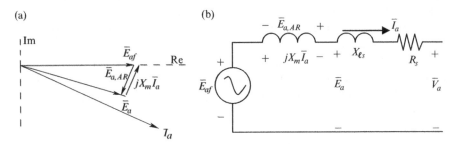

FIGURE 9.10   (a) Phasor diagram; (b) per-phase equivalent circuit.

9.4 POWER OUTPUT, STABILITY, AND THE LOSS OF SYNCHRONISM

Induced emf in stator windings causes phase currents to flow, producing an electromechanical torque that opposes the torque supplied by the turbine. In the circuit shown in Figure 9.11a, consider a generator connected to an infinite bus (an ideal voltage source) $\bar{V}_\infty$ through a radial line.

The reactance X_T is the sum of the generator synchronous reactance and the internal reactance of the utility grid (plus the leakage reactance of transformer[s], if any). Choosing $\bar{V}_\infty$ as the reference phasor (i.e., $\bar{V}_\infty = V_\infty \angle 0$) and neglecting the circuit resistance in comparison to the reactance, the power from the fundamental concepts in Chapter 3 can be written for all three phases as

$$P = 3\frac{E_{af}V_\infty}{X_T}\sin\delta \tag{9.11}$$

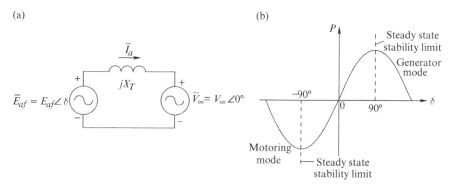

FIGURE 9.11 Power output and synchronism. (a) Infinite bus system; (b) power-angle plot.

where the rotor angle δ associated with $\bar{E}_{af}(= E_{af} \angle \delta)$ is positive in the generator mode; the rotor angle δ is a measure of the angular displacement or position of the rotor with respect to a synchronously rotating reference axis.

If the field current is kept constant, then the magnitude E_{af} is also constant in a steady state, and thus the power output of the generator is proportional to the sine of the torque angle δ between $\bar{E}_{af}$ and $\bar{V}_\infty$. This power-angle relationship is plotted in Figure 9.11b for both positive and negative values of δ, where for negative values of δ, this machine goes into its motor mode.

9.4.1 Steady-State Stability Limit

Figure 9.11b shows that the power supplied by the synchronous generator reaches its peak at $\delta = 90$ degrees. This is the steady-state limit, beyond which the synchronism is lost. This can be explained as follows: initially, assuming no losses, at a value δ_1 below 90 degrees, the turbine is supplying power P_{m1} that equals the electrical output P_{e1}, as shown in Figure 9.12a. To supply more power, the power input from the turbine is increased (for example, by letting more steam into the turbine). This momentarily speeds up the rotor, causing the torque angle δ associated with the rotor-induced voltage $\bar{E}_{af}$ to increase. Finally, a new steady state is reached at $P_{e2}(= P_{m2})$, a higher value of the torque angle δ_2, as shown in Figure 9.12a.

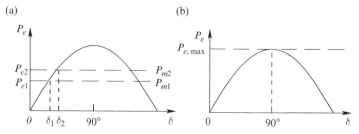

FIGURE 9.12 Steady-state stability limit. (a) Power vs. δ (b) maximum power.

However, at and beyond $\delta = 90$ degrees, if the power input from the turbine is increased, as shown in Figure 9.12b, increasing δ causes the electrical output power to decline, which results in a further increase in δ (because more mechanical power is coming in while less electrical power is going out). This increasing δ causes an intolerable increase in generator currents, and the protection relays cause the circuit breakers to trip to isolate the generator from the grid, thus saving the generator from being damaged.

This sequence of events is called the *loss of synchronism*, and stability is lost. In practice, the transient stability due to a sudden change in the electrical power output forces the maximum value of the steady-state torque angle δ to be much less than 90 degrees, typically in a range of 40 to 45 degrees.

9.5 FIELD EXCITATION CONTROL TO ADJUST REACTIVE POWER

The reactive power associated with synchronous generators can be controlled in magnitude as well as in sign (leading or lagging). To discuss this, let us assume, as a base case, that a synchronous generator is supplying a constant power, and the field current I_f is adjusted such that this power is supplied at a unity power factor, as shown in the phasor diagram in Figure 9.13a.

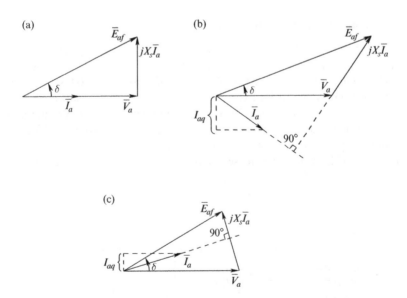

FIGURE 9.13 Excitation control to supply reactive power (neglecting R_s). (a) Phasor diagram with unity power factor; (b) phasor diagram with lagging power factor; (c) phasor diagram with leading power factor.

9.5.1 Overexcitation

An increase in the field current, called *overexcitation*, results in a larger magnitude of $\bar{E}_{af}$, since assuming no magnetic saturation, E_f depends linearly on the field current I_f. However, $E_{af} \sin\delta$ must remain constant from Equation 9.11 since the power output is constant. Similarly, the projection of the current phasor $\bar{I}_a$ on the voltage phasor $\bar{V}_a$ must remain the same as in Figure 9.13a. These result in the phasor diagram in Figure 9.13b, where the current $\bar{I}_a$ is lagging behind $\bar{V}_a$. Considering the utility grid to be a load, the grid absorbs the reactive power as an inductor does. Therefore, a synchronous generator operating in an overexcited mode supplies reactive power, as a capacitor does. The three-phase reactive power Q can be computed from the reactive component of the current I_{aq} as

$$Q = 3V_a I_{aq} \tag{9.12}$$

9.5.2 Underexcitation

In contrast to overexcitation, decreasing I_f results in a smaller magnitude E_{af}, and the corresponding phasor diagram, assuming that the power output remains constant as before, can be represented as in Figure 9.13c. Now the current $\bar{I}_a$ leads the voltage $\bar{V}_a$, and the load (the utility grid) supplies reactive power, as a capacitor does. Thus, a synchronous generator in the underexcited mode absorbs reactive power like an inductor does. The three-phase reactive power Q can be computed from the reactive component of the current I_{aq}, similar to Equation 9.12.

9.5.3 Synchronous Condensers

Sometimes, in power systems, grid-connected synchronous machines are operated in a motor mode to supply reactive power, as shown in Figure 9.14. A similar control over the reactive power in these *synchronous condensers*, as they are often called, can be exercised by controlling the field excitation, as explained previously. There is no need for a turbine to operate these machines, and the small amount of power loss accrued in operating the synchronous machine as a motor is supplied by the grid.

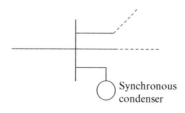

FIGURE 9.14 Synchronous condenser.

9.6 FIELD EXCITERS FOR AUTOMATIC VOLTAGE REGULATION (AVR)

Field excitation of synchronous generators can be controlled to regulate the voltage at their terminals or some other bus in the system, for that matter, usually to its nominal value. This is possible since the voltage regulation and the supply of reactive power are related, and the objective of regulating the voltage at a designated bus dictates what reactive power the generator should supply. Most generators are equipped with an automatic voltage regulator that senses the bus voltage to be regulated and compares it with its desired value. The error between the two is calculated within the regulator, as shown in Figure 9.15, which by means of the phase-controlled rectifier controls the DC voltage applied to the field-excitation winding to adjust the field current I_f appropriately.

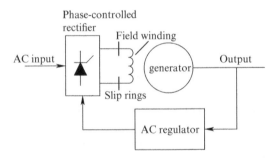

FIGURE 9.15   Field exciter for automatic voltage regulation (AVR).

These field-excitation systems can take several forms based on where the input power is derived from and the desire to avoid slip-rings and brushes required because the field winding to be supplied is rotating with the rotor.

9.7 SYNCHRONOUS, TRANSIENT, AND SUBTRANSIENT REACTANCES

The previous analysis assumes a steady-state operation. However, for example, during and after a short-circuit fault, the rotor oscillations ensue before the rotor reaches another steady state. During these rotor oscillations, the synchronous generator is under a transient condition.

To study transient phenomena, the steady-state per-phase equivalent circuit shown in Figure 9.10b needs to be modified. Modeling synchronous machines can be carried out with increasing levels of complexity, resulting in increased accuracy. However, most fault analyses need an estimate of the fault current. Similarly, in stability studies, usually all that is desired is to determine whether the system would remain stable after a fault and the time required to isolate it. Therefore in these studies, a model called the *constant-flux model* usually suffices, at least for our educational purposes here. A discussion of this model and the resulting equivalent circuit for use in such studies is presented next.

9.7.1 Constant-Flux Model

The field winding of a synchronous generator is supplied by a DC voltage source V_f such that in a steady state, it results in the desired field current $I_f(=V_f / R_f)$, whose DC value is determined by the field resistance R_f. Assuming this DC-excitation voltage to be constant during transients because it cannot change very rapidly, the field winding is essentially a short-circuited coil with a very small resistance relative to its self-inductance L_{ff} and thus has a large time constant. According to the theorem of constant flux linkage, the flux linkage of a short-circuited coil remains constant; that is, it cannot change very quickly. Therefore, under brief transient conditions, as the stator currents suddenly change, causing the armature-reaction flux to change, the field current also suddenly changes to an appropriate value to keep the flux-linkage of the field winding constant.

In a steady state, the armature-reaction flux can penetrate the rotor. Therefore, the steady-state phasor diagram in Figure 9.10a and the per-phase equivalent circuit in Figure 9.10b were obtained using the principle of superposition. Also, the resulting synchronous reactance X_s is large, generally nearly 1 pu on the generator base. The field flux-linkage in the steady state is the sum of the field flux produced by I_f and the armature-reaction flux going through it. At the armature terminal in a steady state, the conditions are $\bar{E}_a$ and $\bar{I}_a$, as shown in Figure 9.10a.

However, immediately following an electrical disturbance, such as a short-circuit fault on the electrical network causing the stator currents to suddenly change, the machine is considered in a subtransient condition. In this subtransient condition, the armature reaction flux cannot enter the field winding due to the constant flux-linkage theorem mentioned earlier, and much of the armature reaction flux is forced to flow through the air gap, resulting in a path of higher magnetic reluctance and hence a smaller reactance. Therefore, the subtransient reactance X_s'' is much smaller than X_s. The ratio between the two can be as large as 4 to 7, where X_s is around 1 pu on the machine base. The armature current and field current, after a sudden three-phase short-circuit at the terminals, are shown in Figures 9.16a and 9.16b, respectively. The subtransient interval corresponds to a few cycles after the fault, where the armature current is very large, and the field current suddenly jumps to maintain the field-flux linkage constant.

A few cycles after the electrical disturbance causes the stator currents to change, but before the steady state, the machine is considered in a transient state. In this condition, some flux lines manage to penetrate the field winding, and the resulting transient reactance X_s' has a value such that $X_s'' < X_s' < X_s$, where X_s' is generally twice of X_s''.

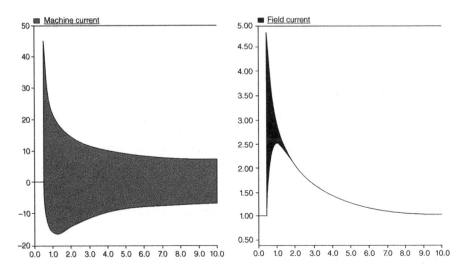

FIGURE 9.16 Armature (a) and field current (b) after a sudden short-circuit. Source: [1].

In simplified models, the saliency effects can be neglected. Therefore, $X_s = X_d$, where X_d is the direct-axis synchronous reactance. Similarly, $X'_s = X'_d$ and $X''_s = X''_d$, where X'_d and X''_d are the direct-axis transient and the direct-axis subtransient reactances, respectively.

To use the subtransient and transient conditions in fault-current calculations and transient-stability studies, respectively, we will modify the steady-state per-phase equivalent circuit in Figure 9.10b for later chapters. We should note that before the fault, the machine is in a steady state, and the terminal conditions $\bar{E}_a$ and $\bar{I}_a$ are as shown in Figure 9.10a, resulting in the per-phase equivalent circuit in Figure 9.10b. Therefore, our model should be such that it results in the appropriate voltage and current at the machine terminal in the steady state and yet is valid in the subtransient and the transient conditions. This can be accomplished in Figure 9.10 by modifying the phasor diagram and per-phase equivalent circuit as shown in Figure 9.17, where

$$\bar{E}_{af} = \bar{E}_a + jX_s\bar{I}_a \quad \bar{E}'_{af} = \bar{E}_a + jX'_s\bar{I}_a \quad \bar{E}''_{af} = \bar{E}_a + jX''_s\bar{I}_a \quad (9.13)$$

Therefore, for a given terminal voltage $\bar{E}_a$ and current $\bar{I}_a$, ignoring the machine resistance, the field-induced voltage can be calculated from Equation 9.13 by using the appropriate reactance based on the type of condition being studied.

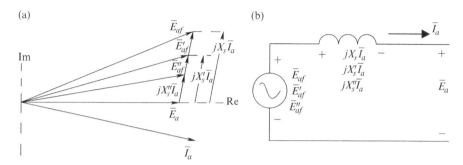

FIGURE 9.17 Synchronous generator modeling for transient and subtransient conditions (in this simplified model, $X_s = X_d$, $X_s' = X_d'$ and $X_s'' = X_d''$).

9.8 GENERATOR MODELING IN PSS®E

9.8.1 Steady-State Model

Generally, a generator is connected to a step-up transformer to increase the voltage before being sent to the grid. There are two ways to model this. We can model it as a separate transformer, or the PSS®E generator model allows us to directly specify the transformer data, as shown in Figure 9.18.

It offers a detailed parameter model, but the primary ones to be defined are the regulating bus number (the default is the self bus where the generator is installed), voltage set-point, and active power generation value to perform the power-flow analysis. The power-flow model representation looks like Figure 9.19.

9.8.2 Dynamic Model

For the dynamic analysis, several models exist in practice. However, as explored later in Chapter 11, a constant internal voltage generator model is used. This model allows us to specify the generator's damping constant and inertia. We will explore other generator models through the homework problems, such as the round rotor generator model (quadratic saturation) and WECC generator model. These models follow a detailed high-order generator model that allows the user to specify d-q synchronous reactance, transient and subtransient reactance, various time constants, and leakage reactance. The Thevenin generator equivalent for switching and dynamic simulation can be represented as shown in Figure 9.20.

Please check the accompanying website for more details on the models and the homework problems.

Machine Data Record   ✕

Power Flow   Short Circuit   NCSFC

Basic Data

Bus Number	1	Bus Name	BUS 1 345.00
Machine ID	1 ☑ In Service	Bus Type Code	3
Baseload Flag	0 - Normal ⌄		
Voltage Droop	None ⌄		

Machine Data   Transformer Data

Pgen (MW)	Pmax (MW)	Pmin (MW)	R Tran (pu)
0.0000	9999.0000	-9999.0000	0.00000
Qgen (Mvar)	Qmax (Mvar)	Qmin (Mvar)	X Tran (pu)
0.0000	9999.0000	-9999.0000	0.00000
Mbase (MVA)	R Source (pu)	X Source (pu)	Gentap (pu)
100.00	0.000000	1.000000	1.00000

Owner Data   Wind Data

Owner		Fraction	Control Mode
1	Select ...	1.000	Conventional Machine ⌄
			Power Factor (WPF)
0	Select ...	1.000	1.000
0	Select ...	1.000	Plant Data
			Sched Voltage
0	Select ...	1.000	1.0000 1

OK   Cancel

FIGURE 9.18   Generator data for the steady-state power flow.

FIGURE 9.19   Generator representation in a steady-state power-flow model.

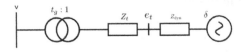

FIGURE 9.20   Generator data for the steady-state power flow.

REFERENCES

1. PSCAD/EMTDC. www.pscad.com.
2. N. Mohan. 2011. *Electric Machines and Drives: A First Course*. John Wiley & Sons.
3. P. Anderson. 1995. *Analysis of Faulted Power Systems*. Wiley-IEEE Press.
4. P. Kundur. 1994. *Power System Stability and Control*. McGraw-Hill.

PROBLEMS

9.1   Assume that the rotor position in Figure 9.8a corresponds to time $t = 0$. Plot e_{af}, e_{bf}, and e_{cf} as functions of $\omega_{syn}t$.

9.2   Assume that the rotor position in Figure 9.8a corresponds to time $t = 0$. Draw the rotor flux-density space vector $\vec{B}_f$ at $\omega_{syn}t$ equal to $0, \pi/6, \pi/3$, and $\pi/2$ radians.

9.3   If the stator phase current angle θ in Equation 9.4 is zero, plot the armature-reaction flux-density space vector $\vec{B}_{AR}$ at $\omega_{syn}t$ equal to $0, \pi/6, \pi/3$, and $\pi/2$ radians.

9.4   In Problem 9.3, plot i_a and $e_{a,AR}$ as functions of $\omega_{syn}t$.

9.5   In the per-phase equivalent circuit in Figure 9.10b, assume $R_s = 0$ and $X_s = 1.2$ pu. The terminal voltage $\bar{V}_a = 1\angle 0$ pu and $\bar{I}_a = 1\angle -\pi/6$ pu . Calculate $\bar{E}_{af}$, and draw a phasor diagram similar to Figure 9.10a.

9.6   In Problem 9.5, with E_{af} kept constant at the magnitude calculated and $V_a = 1$ pu, calculate the maximum power in per-unit that this machine can supply.

9.7   In a synchronous generator, assume $R_s = 0$ and $X_s = 1.2$ pu. The terminal voltage $\bar{V}_a = 1\angle 0$ pu. It supplies 1 pu power. Calculate all the relevant quantities to draw the phasor diagrams in Figure 9.13 if the synchronous generator field-excitation is controlled such that the reactive power Q is as follows: (a) $Q = 0$, (b) supplying $Q = 0.5$ pu, and (c) absorbing $Q = 0.5$ pu.

9.8   Repeat Problem 9.7, assuming that the synchronous machine is a synchronous condenser where the real power $P = 0$.

9.9   In the per-phase equivalent circuit in Figure 9.17b, assume that $R_s = 0$, $X_s = 1.2$ pu, $X'_s = 0.33$ pu, and $X_s = 0.23$ pu. The terminal voltage $\bar{V}_a = 1\angle 0$ pu, and the current $\bar{I}_a = 1\angle -\pi/6$ pu . Draw a phasor diagram similar to Figure 9.17a for steady state, transient, and subtransient operation.

PSS®E-BASED PROBLEMS

9.10   The power flow in a three-bus example power system is calculated using Python and MATLAB in Example 5.4. Repeat this using PSS®E, include a transformer in this example, as described on the accompanying website, and calculate the results of the power-flow study.

9.11   Repeat Example 9.10, where the transformer resistance, reactance, and tap are directly specified in the generator model data instead of a separate transformer, as described on the accompanying website, and calculate the results of the power-flow study.

10

VOLTAGE REGULATION AND STABILITY IN POWER SYSTEMS

10.1 INTRODUCTION

As transmission lines are loaded toward their capacities, voltage stability has become a serious consideration. Several blackouts have been caused by voltage collapse. In this chapter, we will examine the causes of voltage stability, the role of reactive power in maintaining voltage stability, and means to supply reactive power.

10.2 RADIAL SYSTEM AS AN EXAMPLE

To understand the voltage-dependence phenomenon, consider the simple radial system shown in Figure 10.1a, where an ideal source supplies a load through a transmission line of series reactance X_L and shunt susceptances. For simplification, the transmission line resistance is ignored. To analyze such a system, the susceptances on both sides are combined as parts of the sending end and the receiving end systems and represented by the equivalent circuit in Figure 10.1b. To analyze real and reactive powers in the equivalent system in Figure 10.1b, we assume the receiving-end voltage is the reference – that is, $\bar{V}_R = V_R \angle 0$:

$$\bar{I} = \frac{\bar{V}_s - V_R}{jX_L} \tag{10.1}$$

Electric Power Systems with Renewables: Simulations Using PSS®E, Second Edition. Ned Mohan and Swaroop Guggilam.
© 2023 John Wiley & Sons, Inc. Published 2023 by John Wiley & Sons, Inc.
Companion Website: www.wiley.com/go/mohaneps

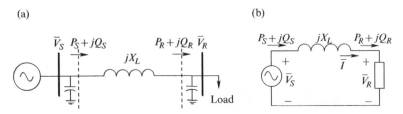

FIGURE 10.1 (a) Simple radial system; (b) equivalent circuit.

At the receiving end, the complex power can be written as

$$S_R = P_R + jQ_R = V_R \bar{I}^*$$ (10.2)

Using the complex conjugate from Equation 10.1 into 10.2 and expressing $\bar{V}_S$ in its polar form as $\bar{V}_S = V_S \angle \delta$,

$$P_R + jQ_R = V_R \left(\frac{V_S \angle(-\delta) - V_R}{-jX_L} \right) = \frac{V_s V_R \sin \delta}{X_L} + j \left(\frac{V_S V_R \cos \delta - V_R^2}{X_L} \right)$$ (10.3)

Equating the real parts on both sides of the equation,

$$P_R = \frac{V_S V_R}{X_L} \sin \delta$$ (10.4)

where, assuming no transmission-line losses, P_R is the same as the sending-end power P_S.
And

$$Q_R = \frac{V_S V_R \cos \delta}{X_L} - \frac{V_R^2}{X_L}$$ (10.5)

Dividing both sides of Equation 10.5 by $\dfrac{V_R^2}{X_L}$ and rearranging terms,

$$\frac{V_R}{V_S} = \cos \delta \left(\frac{1}{1 + \dfrac{Q_R}{V_R^2 / X_L}} \right)$$ (10.6)

In power systems, utilities try to keep bus voltage magnitudes close to their nominal values of 1 pu. Therefore, from Equation 10.4, higher values of the transmission-line loading P_R would require higher values of sin δ and thus lower values of cos δ. Hence, to maintain both voltages close to 1 pu, the reactive power

Q_R must be negative: that is, at higher power loading of a transmission line, the receiving end must supply reactive power locally to maintain its bus voltage.

This requirement for the receiving end to supply reactive power to maintain its voltage can be further explained by the phasor diagram in Figure 10.2a. By Kirchhoff's voltage law in Figure 10.2b,

$$\bar{V}_S = \bar{V}_R + jX_L\bar{I} \tag{10.7}$$

Assuming both bus-voltage magnitudes at 1 pu and $\bar{V}_R$ as the reference voltage, the phasor diagram is shown in Figure 10.2a, where $\bar{V}_S$ leads by an angle δ.

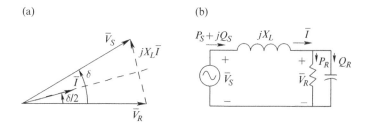

(a)   (b)

FIGURE 10.2   (a) Phasor diagram and (b) equivalent circuit with $V_S = V_R = 1$ pu.

The angle between the two voltages depends on the power transfer over the line, given by Equation 10.4. If both bus voltages are of equal magnitude, $\bar{I}$ from Equation 10.1 is, as shown in Figure 10.2a, exactly at an angle $\delta/2$. This phasor diagram clearly shows $\bar{I}$ in Figure 10.2a leading $\bar{V}_R$, implying that Q_R is negative. This means that to achieve V_R equal to 1 pu voltage, similar to that at the sending end, the receiving end should equivalently appear as shown in Figure 10.2b, where the equivalent resistance absorbs P_R and the equivalent capacitor supplies the reactive power equal to $|Q_R|$. The higher the load, the greater δ and I are, resulting in a higher demand for the reactive power.

It is also useful to know what is happening at the sending end of the transmission line. From the phasor diagram in Figure 10.2a, it can be observed that the sending-end reactive power is the same in magnitude as Q_R but opposite in polarity:

$$Q_S = -Q_R \tag{10.8}$$

Therefore, the sending end supplies reactive power – for example, the generator at the sending end operates overexcited, and the sending-end side susceptance contributes to it as well, to a certain extent.

In the transmission line, the reactive power consumed can be calculated as

$$Q_{\text{Line}} = I^2 X_L \tag{10.9}$$

Just like real power, the reactive power supplied to a system must equal the sum of the reactive power consumed, and thus

$$Q_S = Q_R + \underbrace{I^2 X_L}_{Q_{\text{Line}}} \tag{10.10}$$

Using Equation 10.8 in Equation 10.10, the reactive power consumed by the line is twice $|Q_R|$:

$$I^2 X_L = Q_{\text{Line}} = 2|Q_R| \tag{10.11}$$

In Figure 10.1, the transmission line is represented by lumped elements, and the previous discussion shows what happens at the terminals with which we are mainly concerned. However, the transmission line has distributed parameters, as shown in Figure 10.3a. Assume that the voltages at the two ends are maintained at 1 pu. If this transmission line, assumed to be lossless, is surge-impedance loaded to $P_R = \text{SIL}$, then the voltage profile along the transmission-line length is flat, as shown by the solid line in Figure 10.3b, where the reactive power consumed per unit line length is supplied by its distributed shunt capacitance. Under a heavy load condition with $P_R > \text{SIL}$, the voltage profile will sag, as shown in Figure 10.3b, and both ends must supply reactive power. The opposite is true under light loadings, with $P_R < \text{SIL}$, as shown in Figure 10.3b, where the reactive power supplied by the transmission-line shunt capacitances must be absorbed at both ends to hold voltages at 1 pu.

(a)   (b)

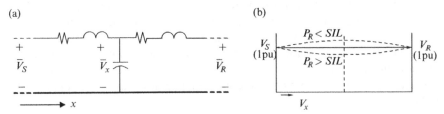

FIGURE 10.3   (a) Transmission line with distributed parameters; (b) voltage profile.

10.3   VOLTAGE COLLAPSE

Once again, let us consider a radial system similar to Figure 10.1b, shown in Figure 10.4a, supplying a load at the receiving end. To begin with, consider a unity power-factor load with $Q_R = 0$.

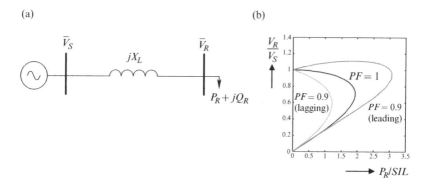

FIGURE 10.4 Voltage collapse in a radial system (example of a 345 kV line, 200 km long). (a) Simple radial system; (b) voltage vs. power.

From Equation 10.5,

$$V_R = V_S \cos \delta \qquad (10.12)$$

Substituting it into Equation 10.4,

$$P_R = \frac{V_S^2}{X_L} \cos \delta \sin \delta \qquad (10.13)$$

To determine the maximum power transfer, taking the partial derivative in Equation 10.13 with respect to angle δ and substituting it to zero results in

$$\frac{\partial P_R}{\partial \delta} = \frac{V_S^2}{X_L} (\cos^2 \delta - \sin^2 \delta) = 0 \qquad (10.14)$$

from which

$$\delta = \pi / 4 \qquad (10.15)$$

Therefore, the maximum power transfer occurs at $\delta = \pi / 4$; and using this condition in Equation 10.13,

$$P_{R,\max} = \frac{V_s^2}{2 X_L} \qquad (10.16)$$

and

$$V_R \simeq 0.7 \, V_s \qquad (10.17)$$

To normalize P_R, it is divided by the surge impedance loading (SIL). With $Q_R = 0$, corresponding to a unity power-factor load, the voltage ratio (V_R / V_S) is plotted in Figure 10.4b as a function of the normalized real power transfer P_R. Similar curves are plotted for loads at lagging and leading power factor of 0:9 for illustration. These "nose" curves show that as the line loading is increased, the receiving-end voltage drops and reaches a critical point that depends on the power factor, beyond which further loading (by reducing the load resistance) at the receiving end results in lower power until the voltage at the receiving end collapses.

A lagging power-factor loading is worse for voltage stability than the unity power-factor loading. From Figure 10.4b, it can be seen that obtaining a near-1 pu receiving-end voltage at a lagging power-factor load would require the sending-end voltage to be unacceptably high.

As shown in Figure 10.4b, leading power-factor loads result in a higher voltage than lagging power-factor loads. However, even at a leading power factor where we would think the receiving-end voltage is higher than normal and the voltage stability should be of no concern, a slight increase in power can lead to the critical point and a possible voltage collapse.

10.4   PREVENTING VOLTAGE INSTABILITY

As the previous analysis shows, voltage instability results from highly loaded systems. Although this has been illustrated using a simple radial system, the same analysis can be applied to a highly integrated system. Voltage instability is associated with the lack of reactive power, and therefore it is necessary to have a reactive power reserve. Several such means are discussed next.

10.4.1   Synchronous Generators

Synchronous generators can supply inductive and capacitive var by excitation control and are the primary source of reactive power. As discussed earlier, heavier loading requires more reactive power support at the receiving and sending ends. However, synchronous generators are limited in how many var they can supply, as shown in Figure 10.5 by a family of curves corresponding to various pressures of hydrogen used for cooling at the rated voltage. A positive value of Q signifies reactive power supplied by the generator in the overexcited mode. Three distinct reactive-power capabilities regions are shown as a function of the real power P that depends on the mechanical input.

In region A, the reactive-power capability is limited by heating due to the armature (stator) current; therefore, the magnitude of the apparent power $|S|(= \sqrt{P^2 + Q^2})$ must not exceed its rated value in a steady state. In region B, the generator is operating overexcited and limited by the field-current heating. In region C, the generator is operating underexcited, and the end-region heating, as explained in [1], can be a problem that limits the armature current. Generally,

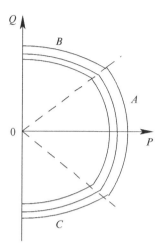

FIGURE 10.5 Reactive power supply capability of synchronous generators.

the intersection of regions *A* and *B* indicates the rating of the synchronous generator in MVA and the power factor at the rated voltage. Curves similar to Figure 10.5 can be plotted for voltages other than 1 pu.

In synchronous generators, the conventional excitation control may be too slow to react; therefore, it is preferable to use a thyristor-based fast-acting excitation control in conjunction with the power system stabilizer (PSS) for damping rotor oscillations. Another possible solution for fast response is to operate the generator overexcited so that normally it produces more var than needed, where extra var are consumed by shunt reactors. Under voltage contingencies, shunt reactors can be quickly disconnected, making those extra var available to the system.

10.4.2 Static Reactive Power Compensators

Recently, power-electronics-based static reactive power-compensating mechanisms have been proposed and implemented for voltage control. These are classified under the category of flexible AC transmission systems (FACTS) [4]. Their operating principles are briefly explained in this section.

The need for reactive power in an area can be met by a shunt device, as shown in Figure 10.6a, where the system we are looking at, including the load at the bus where the device is to be connected, can be represented by its Thevenin equivalent for explanation purposes. The Thevenin impedance is mostly reactive, and it is assumed to be purely so to simplify the explanation. From Figure 10.6a,

$$\bar{V}_{bus} = \bar{V}_{Th} - jX_{Th}\bar{I} \tag{10.18}$$

As the phasor diagrams in Figure 10.6b show, if $\bar{I}$ is leading $\bar{V}_{bus}$, then V_{bus} is higher than V_{Th} due to the voltage drop across the Thevenin reactance; the opposite effect occurs if $\bar{I}$ is lagging $\bar{V}_{bus}$. It is important to recognize from Equation 10.18 that the shunt device affects the bus voltage magnitude by means of the drop across X_{Th}: the smaller the Thevenin reactance, the smaller the effect on the bus voltage. For example, if X_{Th} is nearly zero, no amount of current will affect the bus voltage.

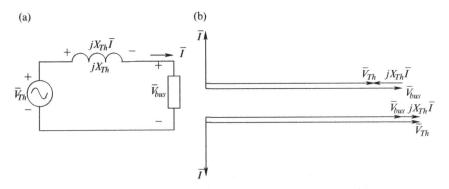

FIGURE 10.6 Effect of leading and lagging currents due to the shunt compensating device. (a) Circuit diagram; (b) phasor diagrams.

The shunt compensating device may consist of capacitor banks that are switched in or out by mechanical means or by back-to-back connected thyristors, as shown in Figure 10.7a. The small inductance shown in the series mainly minimizes transient current at turn-on. The absence of gate pulses to the thyristors keeps them from conducting, whereas applying continuous gate pulses to both thyristors ensures that the current through the pair can flow in either direction as if it were a mechanical switch that is switched on. Often,

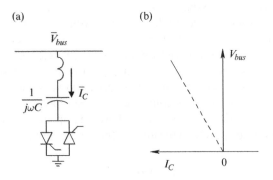

FIGURE 10.7 v–i characteristic of an SVC. (a) SVC connected to the bus; (b) voltage vs. current.

thyristor-switched capacitor banks are referred to as *static var compensators* (SVCs). The *v–i* characteristic of an SVC is a straight line, as shown in Figure 10.7b, where the current magnitude I_C varies linearly with bus voltage as (ωC) V_{bus}.

The shunt-compensating device may consist of a shunt reactor, as shown in Figure 10.8a. Here, the thyristor pair may be designed to act as a switch, as discussed earlier with shunt capacitor banks, switching the reactor in or disconnecting it. However, it is possible to control each thyristor firing angle, as shown by Figure 10.8b, and hence control the current through the reactor, thus controlling its effective reactance and the var provided by it. In this mode, this compensating device is called a *thyristor-controlled reactor* (TCR). At high voltages, the reactor is fully in, with a delay angle of 90 degrees or less. Below a certain threshold voltage, the controller begins to increase the delay angle. When the delay angle reaches 180 degrees, the reactor is completely switched out.

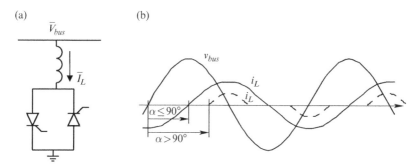

FIGURE 10.8 Thyristor-controlled reactor (TCR).

It is possible to have both SVC and TCR in parallel, as shown in Figure 10.9. By controlling the delay angle of the TCR, the parallel combination can be controlled to either supply or draw reactive power, and its magnitude can also be controlled.

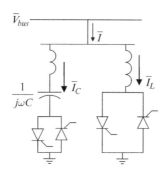

FIGURE 10.9 Parallel combination of SVC and TCR.

10.4.2.1 STATCOMs

In addition to SVCs and TCRs, it is possible to employ static compensators (STATCOMs), which are based on voltage-link converters, as shown in Figure 10.10. As discussed in Chapter 7, dealing with HVDC transmission lines, it is possible to synthesize three-phase sinusoidal voltages from a DC source, such as the DC voltage across the capacitor. A small inductor is at the AC output of the STATCOM, which is not shown in Figure 10.10, mainly for filtering purposes; there is very little drop across it at the fundamental frequency.

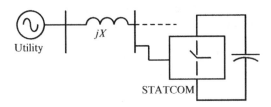

FIGURE 10.10   STATCOM.

In a STATCOM, the line-frequency voltage is synthesized at the output of the converter from the DC-bus voltage across the capacitor; this DC-bus voltage is created and maintained by transferring a small amount of real power from the AC system to the converter to overcome converter losses. Otherwise, there is no real transfer of power through the converter – only controlled reactive power as in an inductor or a capacitor. A STATCOM can draw controllable capacitive or inductive currents independent of the bus voltage. Therefore, a STATCOM can be considered a controllable reactive current source on the bus to which it is connected; its v–i characteristic is shown in Figure 10.11, where the vertical lines represent the current rating of the device.

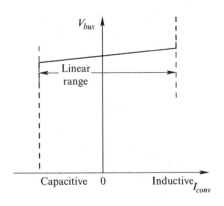

FIGURE 10.11   STATCOM v–i characteristic.

10.4.2.2 STATCOM in PSS®E

A STATCOM is a fast-acting device that regulates the voltage of the bus by supplying or absorbing reactive power at the point of interconnection with the grid without needing external capacitor banks. When the current leads the voltage in the capacitive region, the STATCOM injects reactive power to increase the bus voltage. When the current lags the voltage in the inductive region, the STATCOM absorbs reactive power to reduce the bus voltage. As shown in Figure 10.11, the current can be capped at a specified max value when the voltage is outside the linear region, which gives this an edge over static var compensators. For the steady-state power flow application, a STATCOM is represented in PSS®E as a FACTS device model that allows voltage regulation at the specified bus.

Example 10.1

In Example 5.4, add a STATCOM device regulating the bus 3 voltage at 1.02 pu, and set the current limit to 500 MVA.

Solution The results are graphically shown in Figure 10.12, and the case files can be found on the accompanying website.

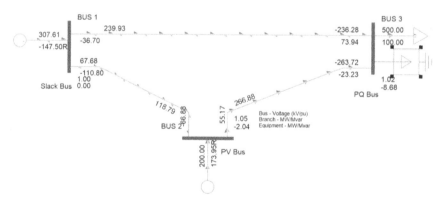

FIGURE 10.12 Power-flow model for a three-bus system with a STATCOM at bus 3 in PSS®E.

10.4.3 HVDC-VSC Systems

If the power transfer between two systems or areas takes place using a voltage-link HVDC line, the converters on both sides of a HVDC-VSC line can independently supply or absorb reactive power as needed.

10.4.4 Thyristor-Controlled Series Capacitor (TCSC)

Other means for voltage support are series capacitors, which reduce the effective value of X_L in the power equation in Equation 10.4. In addition to series capacitors, it is possible to insert a thyristor-controlled series capacitor (TCSC), as shown in Figure 10.13.

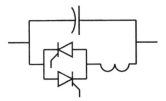

FIGURE 10.13   Thyristor-controlled series capacitor (TCSC) [4].

In TCSCs, the effective inductance of the inductor, in parallel with the capacitor, can be controlled by the conduction angles of the thyristors, thus controlling the effective reactance of the TCSC to appear as capacitive or inductive. One such unit operates in the western part of the United States [5].

10.4.5   Unified Power-Flow Controller (UPFC) and Static Phase Angle Control

In addition to the FACTS devices mentioned earlier, additional devices can directly or indirectly help with voltage stability. Based on Equation 10.4, a device connected to a bus in a substation, as shown in Figure 10.14a, can influence power flow in three ways:

1. Controlling the voltage magnitudes
2. Changing the line reactance and/or X
3. Changing the power angle δ

One such device, called a unified power flow controller (UPFC) [4], can affect power flow in any combination of these ways. A block diagram of a UPFC is shown in Figure 10.14a at one side of the transmission line. It consists of two voltage-source switch-mode converters. The first converter injects a voltage $\bar{E}_3$ in series with the phase voltage such that

$$\bar{E}_1 + \bar{E}_3 = \bar{E}_2 \qquad (10.19)$$

Therefore, by controlling the magnitude and phase of the injected voltage $\bar{E}_3$ within the circle shown in Figure 10.14b, the magnitude and phase of the bus voltage $\bar{E}_2$ can be controlled. If a component of the injected voltage $\bar{E}_3$ is made to be 90 degrees out-of-phase – for example, leading with respect to the current phasor $\bar{I}$ – then the transmission line reactance X is partially compensated.

The second converter in a UPFC is needed for the following reason: since converter 1 injects a series voltage $\bar{E}_3$, it delivers real power P_1 and reactive power Q_1 to the transmission line (where P_1 and Q_1 can be either positive or negative):

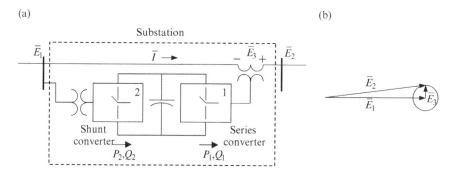

FIGURE 10.14 Unified power flow controller (UPFC).

$$P_1 = 3\,\mathrm{Re}(\bar{E}_3 \bar{I}^*) \tag{10.20}$$

$$Q_1 = 3\,\mathrm{Im}(\bar{E}_3 \bar{I}^*) \tag{10.21}$$

Since there is no steady-state energy storage capability within the UPFC, the power P_2 into converter 2 must equal P_1 if the losses are ignored:

$$P_2 = P_1 \tag{10.22}$$

However, the reactive power Q_2 bears no relation to Q_1 and can be independently controlled within the voltage and current ratings of converter 2:

$$Q_2 \neq Q_1 \tag{10.23}$$

By controlling Q_2 to control the magnitude of the bus voltage $\bar{E}_1$, a UPFC provides the same functionality as an advanced static var compensator STATCOM. A UPFC combines several functions: static var compensator, phase-shifting transformer, and controlled series compensation.

REFERENCES

1. P. Kundur. 1994. *Power System Stability and Control*. McGraw-Hill.
2. C. W. Taylor. 1994. *Power System Voltage Stability*. McGraw-Hill. (For reprints, email cwtaylor@ieee.org.)
3. *PowerWorld* Computer Program. www.powerworld.com.
4. N. Hingorani and L. Gyugyi. 1999. *Understanding FACTS: Concepts and Technology of Flexible AC Transmission Systems*. Wiley-IEEE Press.
5. W. Breuer, D. Povh, D. Retzmann, Ch. Urbanke, and M. Weinhold. "Prospects of Smart Grid Technologies for a Sustainable and Secure Power Supply." The 20th World Energy Congress and Exposition, Rome, Italy, November 11–25, 2007.

PROBLEMS

10.1   In the power-flow example in Chapter 5, in the three-bus power system, what reactive power compensation needs to be provided at bus 3 to bring its voltage to 1 pu?

10.2   In the power flow example in Chapter 5, in the three-bus power system, what will the voltage at bus 3 be if the power demand at bus 3 is reduced by 50%?

10.3   Based on the voltage sensitivity to a reactive power change at bus 3, as calculated in Chapter 5 for the example power system, what reactive power is needed at bus 3 to bring its voltage to 1 pu? Compare this result with that of Problem 10.1.

10.4   Calculate the reactive power consumed by the three transmission lines in the example power system in Chapter 5.

10.5   If the reactive power compensation in Problem 10.1 is provided by shunt capacitors, calculate their value.

10.6   If the reactive power compensation in Problem 10.1 is provided by a STATCOM, calculate the equivalent $\overline{V}_{conv}$ in pu; assume X as shown in Figure 10.10 to be 0.01 pu.

10.7   In the power-flow example in Chapter 5, what will the effect be on the bus 3 voltage if lines 1–3 and 2–3 are 50% compensated by series capacitors?

10.8   In the power-flow example in Chapter 5, what must the voltage be at bus 1 to bring the bus 3 voltage to 1 pu?

10.9   Why is the performance of STATCOMs superior to that of shunt capacitors?

PSS®E-BASED PROBLEMS

10.10   Confirm the results of Problem 10.1.
10.11   Confirm the results of Problem 10.2.
10.12   Confirm the results of Problem 10.4.
10.13   Confirm the results of Problem 10.5.
10.14   Confirm the results of Problem 10.6.
10.15   Confirm the results of Problem 10.7.
10.16   Confirm the results of Problem 10.8.
10.17   In Example 10.1, limit the max current to 50 MVA, re-run the power-flow analysis, and compare the results.

11

TRANSIENT AND DYNAMIC STABILITY OF POWER SYSTEMS

11.1   INTRODUCTION

In an interconnected power system such as that in North America, thousands of generators operate in synchronism with each other. They share load based on economic dispatch and optimum power flow, as discussed in Chapter 12. But major disturbances, even momentarily, such as a fault in the system, loss of generation, or sudden loss of load, can threaten this synchronous operation. The ability of the power system to maintain synchronism when subjected to this kind of major disturbance is called *transient stability*. Such an interconnected system should also have enough damping to retain dynamic stability, as explained later in this chapter.

11.2   PRINCIPLE OF TRANSIENT STABILITY

The principle of transient stability can be illustrated by a simple system with one generator connected through a transformer to an infinite bus, considered an ideal voltage source with a voltage of $\bar{V}_B (= V_B \angle 0)$, as shown in Figure 11.1a by two parallel lines.

Electric Power Systems with Renewables: Simulations Using PSS®E, Second Edition. Ned Mohan and Swaroop Guggilam.
© 2023 John Wiley & Sons, Inc. Published 2023 by John Wiley & Sons, Inc.
Companion Website: www.wiley.com/go/mohaneps

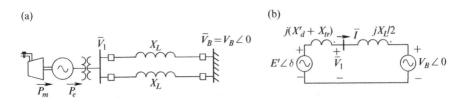

FIGURE 11.1 Simple one-generator system connected to an infinite bus. (a) One line diagram; (b) circuit representation.

The power delivered by the mechanical source is P_m, which equals the generator electrical output P_e in a steady state, assuming all the generator losses to be zero. As discussed in Chapter 9, under transient conditions, using the constant flux model, a synchronous generator can be represented by a voltage source of a constant amplitude at the back of the transient reactance X'_d of the generator, as shown in Figure 11.1b. Here, $\bar{E}' (= E' \angle \delta)$ is such that in a steady state, before the fault, the equivalent circuit in Figure 11.1b results in $P_e = P_m$. In Figure 11.1b, X_{tr} is the leakage reactance of the transformer. Ignoring all the losses in the system in Figure 11.1b, as derived in Chapter 2, the electrical power in MW delivered by the generator to the infinite bus is

$$P_e = \frac{E'V_B}{X_{T1}}\sin\delta \qquad (11.1)$$

where E' and V_B are the magnitudes of the two voltages in kV that are at an angle δ (in electrical radians) apart and connected through the total reactance X_{T1} (in Ω), which is the sum of the generator transient reactance X'_d, the transformer leakage reactance X_{tr}, and the reactance of the two transmission lines in parallel, $X_L/2$. If this transient stability study lasts a second or less, it is reasonable to assume as a first-order approximation that the exciter system of the generator cannot respond in such a short time. Therefore, the magnitude of the voltage behind the transient reactance can be assumed constant. Similarly, the turbine governor control, regardless of whether it is a steam or hydro turbine, cannot react within this short time. Hence, the mechanical power input P_m from the turbine to the generator can be assumed constant.

In a steady state, the generator rotates at a mechanical speed ω_m (in mechanical radians/s), which equals the synchronous speed. The generator's electrical power output, P_e, based on Equation 11.1, is plotted in Figure 11.2a as a function of the rotor angle δ (in electrical radians) associated with the generator voltage. In a steady state before the disturbance, $P_e = P_m$, and the initial rotor angle is δ_0, as shown in Figure 11.2a.

If a fault occurs – for example, on one of the transmission lines, as shown in Figure 11.2b – then during the fault, the ability to transfer electrical power P_e, shown by the dashed-dotted curve in Figure 11.2a, goes down because the bus 1 voltage goes down. It takes a finite clearing time for the circuit breakers at both

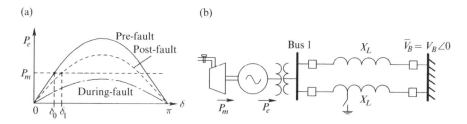

FIGURE 11.2 (a) Power angle curve; (b) single line diagram with a fault on a transmission line.

ends to isolate the faulted line; after the fault-clearing time, the generator and the infinite bus are connected by the remaining transmission line. Using an equation similar to Equation 11.1, the power-angle curve for the post-fault curve is shown dotted in Figure 11.2a. It is clear from Figure 11.2a that if the system is stable after the faulted line is taken out of service, the new steady-state rotor-angle value will be δ_1. In this section, we will look at the dynamics of how the rotor angle reaches this new steady state, assuming that the transient stability is maintained.

11.2.1 Rotor-Angle Swing

During and following a disturbance, until a new steady state is reached, $P_e \neq P_m$. As explained in Appendix 11A, this results in an electrical torque T_e that does not equal the mechanical torque T_m, and the difference of these torques causes the rotor speed ω_m to deviate slightly from the synchronous speed. Therefore, the rotor angle δ_m in mechanical radians per second can be described as

$$J_m \frac{d^2\delta_m}{dt^2} = T_m - T_e \tag{11.2}$$

where J_m is the moment of inertia of the rotational system, which acts upon the acceleration torque, which is the difference between the mechanical torque input T_m and the electrical generator torque T_e opposing it. As described in Chapter 9, note that the rotor angle δ_m is a measure of the angular displacement or position of the rotor with respect to a synchronously rotating reference axis. Multiplying both sides of Equation 11.2 by the rotor mechanical speed ω_m, and recognizing that the product of the torque and speed equals power, Equation 11.2 can be written as

$$\omega_m J_m \frac{d^2\delta_m}{dt^2} = P_m - P_e \tag{11.3}$$

To express this equation in per-unit, a new inertia-related parameter H_{gen} is defined whose value lies in a narrow range of 3 to 11 s for turbo-alternators and 1 to 2 s for hydrogenerators. It is defined as the ratio of the kinetic energy of the rotating mass at the rated synchronous speed $\omega_{syn,m}$ in mechanical radians/s to the three-phase volt-ampere rating $S_{rated,gen}$ of the generator:

$$H_{\text{gen}} = \frac{\frac{1}{2}J_m\omega_{\text{syn},m}^2}{S_{\text{rated,gen}}} \tag{11.4}$$

Substituting for J_m in terms of H_{gen}, defined in Equation 11.4, Equation 11.3 can be written as

$$\left(\frac{\omega_m}{\omega_{\text{syn},m}^2}\right)2H_{\text{gen}}\frac{d^2\delta_m}{dt^2} = P_{m,\text{gen,pu}} - P_{e,\text{gen,pu}} \tag{11.5}$$

where P_m and P_e are in per-unit of the generator MVA base $S_{\text{rated,gen}}$. Often, the system MVA base S_{system} is chosen to be 100 MVA. Regardless of the chosen value of S_{system}, in per-unit of the system MVA base, Equation 11.5 can be written as

$$\left(\frac{\omega_m}{\omega_{\text{syn},m}^2}\right)2H\frac{d^2\delta_m}{dt^2} = P_{m,\text{pu}} - P_{e,\text{pu}} \tag{11.6}$$

where $P_{m,\ pu}$ and $P_{e,\ pu}$ are in per-unit of the system MVA base S_{system} and

$$H = H_{\text{gen}}\left(\frac{S_{\text{rated,gen}}}{S_{\text{system}}}\right) \tag{11.7}$$

Note that even under the transient condition, it is reasonable to assume in Equation 11.6 that the rotor mechanical speed ω_m is approximately equal to the synchronous speed corresponding to the frequency of the infinite bus: $\omega_m \approx \omega_{\text{syn},m}$. Therefore, Equation 11.6 can be written as

$$\frac{2H}{\omega_{\text{syn},m}}\frac{d^2\delta_m}{dt^2} = P_{m,\text{pu}} - P_{e,\text{pu}} \tag{11.8}$$

In Equation 11.8, both the angle deviation and the synchronous speed can be expressed in terms of electrical radians:

$$\frac{2H}{\omega_{\text{syn}}}\frac{d^2\delta}{dt^2} = P_{m,\text{pu}} - P_{e,\text{pu}} \tag{11.9}$$

This equation is called the *swing equation* and describes how the angle δ swings or oscillates due to an imbalance between the mechanical power input and the electrical power output of the generator. To calculate the dynamics of the speed and angle as functions of time, several sophisticated numerical methods can be used. But to illustrate the basic principle, we will use Euler's method, which is the simplest. We will assume that the integration with time is carried out with a sufficiently small time increment Δt during which $(P_m - P_e)$ can be assumed constant. With this assumption, integrating both sides of Equation 11.9 with time,

$$\left.\frac{d\delta}{dt}\right|_t = \omega(t) = \omega(t - \Delta t) + \frac{\omega_{syn}}{2H}(P_{m,pu} - P_{e,pu})\Delta t \qquad (11.10)$$

Similarly, assuming the speed in Equation 11.10 to be constant at $\omega(t - \Delta t)$ during Δt,

$$\delta(t) = \delta(t - \Delta t) + \omega(t - \Delta t)\Delta t \qquad (11.11)$$

Equations 11.10 and 11.11 show the rotor dynamics in terms of its speed ω and the angle δ as functions of time. Example 11.1 shows the rotor dynamics in a simple system with one generator connected to an infinite bus.

Example 11.1
Consider the simple system discussed earlier in Figure 11.2b. The infinite-bus voltage is $\bar{V}_B = 1\angle 0$ pu. The bus 1 voltage magnitude is $V_1 = 1.05$ pu. The generator has a transient reactance $X'_d = 0.28$ pu at a base of 22 kV (L-L) and its three-phase 1,500 MVA base. On the generator base, $H_{gen} = 3.5$ s. The transformer steps up 22 kV to 345 kV and has a leakage reactance of $X_{tr} = 0.2$ pu at its base of 1,500 MVA. The two 345 kV transmission lines are 100 km in length, and each has a series reactance of 0.367 Ω/km, where the series resistance and shunt capacitances are neglected. Initially, the three-phase power flow from the generator to the infinite bus is 15 MW. The system MVA base is 100.

A three-phase to ground fault occurs on one of the lines from bus 1. Calculate the maximum rotor-angle swing δ_m if the fault clearing is 40 ms, after which the faulted transmission line is isolated from the system by the circuit breakers at both ends of the line.

Solution This example is solved using a program written in Python and MATLAB, which is verified by solving it in PSS®E [1]. Both of these are described on the accompanying website. The plot of the rotor-angle oscillation is shown in Figure 11.3; these oscillations continue since no damping is included in the model.

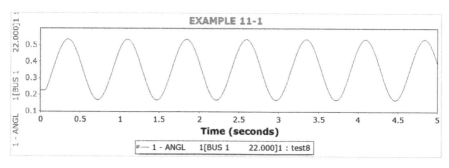

FIGURE 11.3 Rotor oscillation in Example 11.1 from PSS®E.

11.2.2 Determining Transient Stability Using an Equal-Area Criterion

Consider the simple system discussed earlier, repeated in Figure 11.4a, where a fault occurs on one of the transmission lines and is cleared after a time t_{cl} by isolating the faulted line from the system. In the steady state before the fault, $d\delta/dt = 0$ at δ_0; this is the initial steady-state value of δ in Figure 11.4b, given by the intersection of the horizontal line representing P_m and the pre-fault curve.

(a)   (b)

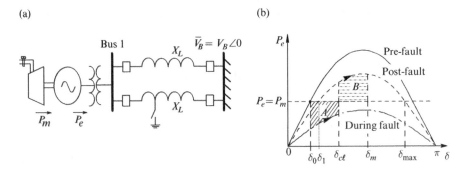

FIGURE 11.4 (a) Single-line diagram with a fault on one transmission line; (b) power angle curves.

The behavior of the rotor angle can be determined by multiplying both sides of Equation 11.9 by $d\delta/dt$:

$$2\frac{d\delta}{dt}\frac{d^2\delta}{dt^2} = \frac{\omega_{\text{syn}}}{H}(P_{m,\text{pu}} - P_{e,\text{pu}})\frac{d\delta}{dt} \tag{11.12}$$

Replacing δ in this equation with θ and integrating both sides with respect to time, we get

$$\int\left(2\frac{d\theta}{dt}\frac{d^2\theta}{dt^2}\right)dt = \frac{\omega_{\text{syn}}}{H}\int(P_{m,\text{pu}} - P_{e,\text{pu}})\frac{d\theta}{dt}dt \tag{11.13}$$

where the integral on the left side equals $(d\theta/dt)^2$. Therefore, in Equation 11.13, from the initial angle δ_0 at which $d\delta/dt = 0$ to an arbitrary angle δ,

$$\left(\frac{d\delta}{dt}\right)^2 = \frac{\omega_{\text{syn}}}{H}\int_{\delta_0}^{\delta}(P_{m,\text{pu}} - P_{e,\text{pu}})d\delta \tag{11.14}$$

In a stable system, the maximum value of the rotor angle will reach some value δ_m, as shown in Figure 11.4b, at which instant $d\delta/dt$ once again becomes zero and the value of δ begins to decrease. Substituting this condition in Equation 11.14,

$$\frac{\omega_{\text{syn}}}{H}\int_{\delta_0}^{\delta_m}(P_{m,\text{pu}} - P_{e,\text{pu}})d\delta = 0 \tag{11.15}$$

In Equation 11.15 and Figure 11.4b, $P_e = P_{e,\,\text{fault}}$ during the fault duration, with $\delta_0 < \delta < \delta_{cl}$ and $P_e = P_{e,\text{post-fault}}$ after the fault is cleared with $\delta_{cl} \leq \delta \leq \delta_m$. Therefore, Equation 11.15 can be written as

$$\underbrace{\int_{\delta_0}^{\delta_c}(P_{m,\text{pu}} - P_{e,\text{fault,pu}})d\delta}_{\text{Area } A} - \underbrace{\int_{\delta_c}^{\delta_m}(P_{e,\text{post-fault,pu}} - P_{m,\text{pu}})d\delta}_{\text{Area } B} = 0 \tag{11.16}$$

which shows that in a stable system, area A equals area B in magnitude.

While the faulted line is still connected to the system, $d\delta/dt$ given by Equation 11.14 is positive in Figure 11.4b since the mechanical power input exceeds the electrical power output and the rotor angle increases from δ_0 to a new value δ_{cl} at the clearing-time t_{cl}. area A, given by Equation 11.16 and illustrated graphically in Figure 11.4b, represents the excess energy delivered to the inertia of the rotor, causing the rotor angle to increase.

After the clearing time t_{cl}, at which time the rotor angle has reached δ_{cl}, circuit breakers at both ends of the faulted transmission line open to isolate the faulted line from the rest of the system and the electrical output shifts to the post-fault power-angle curve, shown dotted in Figure 11.4b. Now the electrical power output exceeds the mechanical power input in the second part of the integral in Equation 11.16. Therefore, beyond δ_{cl} in Figure 11.4b, the speed begins to decrease (although it is still above the synchronous speed), causing the angle to keep increasing, as shown in Figure 11.4b. When area A equals area B in magnitude, the excess energy supplied to the rotor equals that given up by it, and the rotor speed returns to its original synchronous speed value. At this time, the rotor angle reaches its maximum value δ_m, $d\delta/dt = 0$ in Equation 11.14, and area B equals area A.

Thus the equal-area criterion given by Equation 11.16 is able to determine the maximum swing of the rotor angle. To maintain stability, δ_m, determined by the equal-area criterion in Equation 11.16, must be less than $\delta_{\max}$, as labeled in Figure 11.4b, for the generator to remain in synchronism, as further explained in Section 11.2.2.1. After the rotor angle reaches δ_m, the electrical output in Figure 11.4a is still greater than the mechanical input; hence the speed begins to decline and the rotor angle δ begins to decrease, as shown in Figure 11.5.

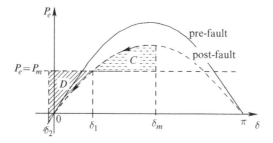

FIGURE 11.5 Rotor oscillations after the fault is cleared.

With the lack of damping assumed here, the angle would oscillate forever around δ_1 in Figure 11.5, between δ_m and δ_2, with areas C and D equal in magnitude. However, the damping in a real system would eventually cause the rotor angle to settle down at δ_1.

11.2.2.1 Critical Clearing Angle

Consider the case shown in Figure 11.6, where the clearing time is larger than before, such that with areas A and B equal in magnitude, δ_m is equal to δ_{max}.

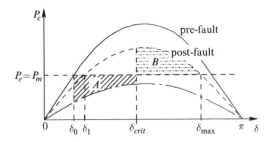

FIGURE 11.6 Critical clearing angle.

Figure 11.6 represents the limiting case of the critical (maximum) clearing time and hence δ_{crit} at that time, beyond which the stability would be lost. Prior to the rotor angle reaching δ_{max}, if the generator inertia cannot give up the excess energy acquired during the fault period, it will not be able to slow down. The reason is that beyond δ_{max}, the mechanical power input is greater than the electrical power output under the post-fault condition. Therefore, the rotor angle will keep increasing, resulting in *out-of-step operation*. This causes the relays to trip the circuit breakers and isolate the generator to prevent damage from excessive currents in the system, and stability is lost. For a given type and location of a fault or a sudden change in the electrical load, there is a critical angle δ_{crit} corresponding to a maximum critical clearing time t_{crit} that results in area A equal to area B in magnitude, as shown in Figure 11.6. Note that $\delta_{max} = \pi - \delta_1$.

Of course, this discussion is theoretical. In practice, there must be a sufficient safety margin to maintain transient stability.

Example 11.2

Consider the simple system discussed earlier in Figure 11.4a. The infinite-bus voltage is $\overline{V}_B = 1\angle 0$ pu. The voltage magnitude at bus 1 is $V_1 = 1.05$ pu. The generator has a transient reactance $X_d' = 0.28$ pu at a base of 22 kV (L-L) and 1,500 MVA. On the generator base, $H_{gen} = 3.5$ s. The transformer steps up 22 kV to 345 kV and has a leakage reactance of $X_{tr} = 0.2$ pu at its own base of 1,500 MVA. The two 345 kV transmission lines are 100 km in length, and each has a series reactance of 0.367 Ω/km, where the series resistance and the shunt capacitances are neglected. Initially, the three-phase power flow from the generator to the infinite bus is 1500 MW.

A three-phase to ground fault occurs on one of the lines, 20% of the distance from bus 1. Calculate the maximum rotor-angle swing δ_m if the rotor angle at the time of fault clearing is 50 degrees.

Solution The solution to this example is carried out by a Python program included on the accompanying website. The power angle curves for the pre-fault, during-fault, and post-fault conditions are shown in Figure 11.7, where initially $\delta_0 = 33.50$ degrees. The peak values of the power-angle curves are calculated as follows on a system MVA base of 100 MVA: $\hat{P}_{e,\text{pre-fault,pu}} = 27.17$, $\hat{P}_{e,\text{fault,pu}} = 5.78$, and $\hat{P}_{e,\text{post-fault,pu}} = 20.51$. Using these values, the power-angle curves are as shown in Figure 11.7.

During the fault, area A in Figure 11.7 can be calculated using Equation 11.16:

$$\text{Area } A = \int_{\delta_{cl}}^{\delta_0} (P_{m,\text{pu}} - \hat{P}_{e,\text{fault,pu}} \sin\delta)d\delta \tag{11.17}$$

$$= P_{m,\text{pu}}(\delta_{cl} - \delta_0) + \hat{P}_{e,\text{fault,pu}}(\cos\delta_{cl} - \cos\delta_0)$$

Similarly, area B in Figure 11.7 can be calculated from Equation 11.16:

$$\text{Area } B = \int_{\delta_m}^{\delta_{cl}} (\hat{P}_{e,\text{post-fault,pu}} \sin\delta - P_{m,\text{pu}})d\delta \tag{11.18}$$

$$= \hat{P}_{e,\text{post-fault,pu}}(\cos\delta_{cl} - \cos\delta_m) - P_{m,\text{pu}}(\delta_m - \delta_{cl})$$

By applying the equal-area criterion, the maximum rotor-angle swing $\delta_m = 95.27$ degrees, as shown in Figure 11.7.

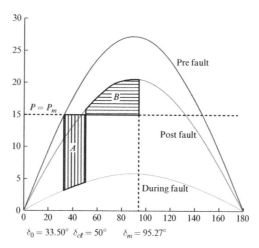

FIGURE 11.7 Power-angle curves and equal-area criterion in Example 11.2.

11.3   TRANSIENT STABILITY EVALUATION IN LARGE SYSTEMS

The equal-area method describes the principle behind transient stability. However, in practice, transient stability must be evaluated in the presence of a large number of generators. In such a system, shown in Figure 11.8 as a block diagram, the rotor's electromechanical dynamics are represented in the time domain, and the electrical network is represented in the phasor domain.

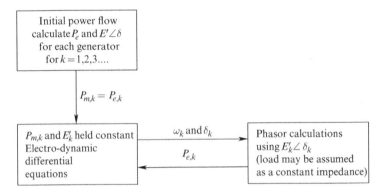

FIGURE 11.8   Block diagram of a transient stability program for an *n*-generator case.

The role of electromechanical dynamics is to provide the rotor angle (angle of synchronous generator voltages), whereas the role of network phasor calculations is to calculate the electrical power that various generators provide at that time. Phasor calculations are made assuming that the network is in a quasi-steady state. This way, the line frequency in the system is eliminated. This procedure, illustrated in the block diagram in Figure 11.8, allows large time steps to be taken based on the electromechanical time constants of the system.

It is important that the entire network not be calculated in the time domain, which would require a small simulation time step and a prohibitively large execution step. A time-domain procedure needs to be adopted using a program such as EMTDC only if a certain phenomenon, such as the performance of a thyristor-controlled series capacitor (TCSC), is being evaluated. The procedure for stability analysis of large networks is illustrated using a three-bus system in Example 11.3.

Example 11.3

Consider the three-bus system discussed in Chapter 5 and repeated in Figure 11.9. These three buses are connected through three 345 kV transmission lines 200, 150, and 150 km long, as shown in Figure 5.1 in Chapter 5.

These transmission lines, considered to consist of bundled conductors, have a line reactance of 0.367 Ω/km at 60 Hz. The line resistance is 0.0367 Ω/km. Ignore all the shunt susceptances. Bus 1 is a slack bus with $V_1 = 1.0$ pu and $\theta_1 = 0$. Bus 2 is a PV bus with $V_2 = 1.05$ pu and $P_2^{sp} = 4.0$ pu. Bus 3 is a PQ bus with injection $P_3^{sp} = -5.0$ pu and $Q_3^{sp} = -1.0$ pu.

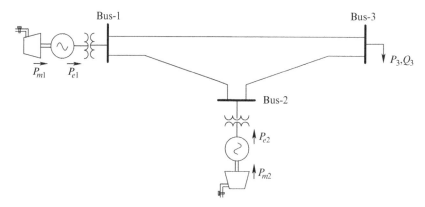

FIGURE 11.9 A 345 kV test system for Example 11.3.

Both the transformers and the generators have three-phase MVA ratings of 500 MVA each. Both generators have a transient reactance $X_d' = 0.23$ pu at a base of 22 kV (L-L) and their own MVA base. Also, each generator has $H_{gen} = 3.5$ s on the generator base. Each 22-to-345 kV step-up transformer has a leakage reactance of $X_{tr} = 0.2$ pu at its own MVA base.

A three-phase to ground fault occurs on line 1–2, one-third of the distance from bus 1. Calculate the rotor-angle swings if the fault-clearing time is 0.1 s, after which the faulted transmission line is isolated from the system by the circuit breakers at both ends of line 1–2.

Solution The solution to this example using Python, MATLAB, and PSS®E [1] is included on the accompanying website.

11.4 DYNAMIC STABILITY

Without adequate damping, an interconnected power system can develop growing power oscillations [2], causing it to split and possibly resulting in a blackout. One such incident took place in the western USA/Canada system on 10 August 1996, as shown by the plot of power oscillations in Figure 11.10, and left millions of customers without power.

This incident, among many, illustrates that there should be adequate damping in the system so that oscillations caused by changes in loads, regardless of the size of these changes, decay to a new steady-state operating point. This is

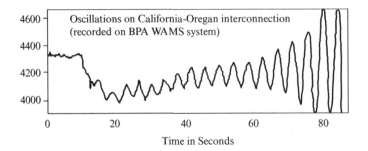

FIGURE 11.10 Growing power oscillations: western USA/Canada system, 10 August 1996 [3].

considered *dynamic stability*, which has become particularly important in modern power systems with fast-acting feedback controllers used in excitation systems, HVDC transmission systems, and FACTS (flexible AC transmission systems [4]) devices. With proper design, for example, using power system stabilizers in conjunction with exciters for synchronous generators and HVDC control [5], it is possible to provide the required damping to maintain dynamic stability. Series-capacitor compensation of transmission lines can lead to sub-synchronous resonances that can fatigue the turbine-generator shaft and therefore must be damped out. To complete the dynamic stability investigation and design the controllers accordingly, a modal analysis is required that uses the concept of eigenvectors. Therefore, as important as this topic is, it is beyond the scope of the first course on power systems.

PSS®E provides tools to perform various dynamic simulations that can help determine the system's transient stability. The PSS®E approach is to take the given case and convert the generators to their Norton equivalents to represent the constant current injection model. After more conversions, a sequence of steps is followed to apply a fault, perform dynamic simulation, clear the fault, continue the dynamic simulation, and eventually analyze the results. More detailed steps about this approach can be found on the accompanying website, along with the simulation files for the specific examples.

REFERENCES

1. PowerWorld Computer Program. www.powerworld.com.
2. G. Rogers. 2000. *Power System Oscillations*. Kluwer.
3. J. F. Hauer et al. 2000. "Dynamic Performance Validation in the Western Power System." APEx 2000, Kananaskis, Alberta, Canada. http://certs.lbl.gov/pdf/apex2000.pdf.
4. N. Hingorani and L. Gyugyi. 1999. *Understanding FACTS: Concepts and Technology of Flexible AC Transmission Systems*. Wiley-IEEE Press.
5. R. L. Cresap et al. 1978. "Operating Experience with Modulation of the Pacific HVDC Intertie." *IEEE Transactions on PA&S*. Vol. PAS-97, pp. 1053–1059.

PROBLEMS

11.1   Redo Example 11.1 with a fault-clearing time that is twice as long.

11.2   In Example 11.2, what is the value of the voltage E', as defined in Figure 11.1b?

11.3   In Example 11.2, what is the time taken by the rotor angle to reach $\delta_{cl} = 75$ degrees, at which time the fault is cleared?

11.4   In Example 11.2, what is the value of δ_{crit}, and what is the time taken by the rotor angle to reach that value?

11.5   Redo Example 11.3 if the initial real and reactive powers at bus 3 are 75% of their original values.

11.6   Redo Example 11.3 if the generators H_{gen} are 50% of their original value.

11.7   Redo Example 11.3 if a fault occurs on line 1–3, one-third of the distance from bus 1.

APPENDIX 11A   INERTIA, TORQUE, AND ACCELERATION IN ROTATING SYSTEMS

Synchronous generators are a rotating type. Consider a lever, pivoted and free to move as shown in Figure 11A.1. When an external force f is applied in a *perpendicular* direction at a radius r from the pivot, then the torque acting on the lever is

$$\underset{[Nm]}{T} = \underset{[N]}{f} \; \underset{[m]}{r} \tag{11A.1}$$

which acts in a counterclockwise direction, considered here to be positive.

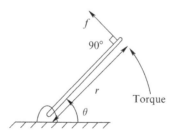

FIGURE 11A.1   Pivoted lever.

    In a turbine-generator system, forces shown by arrows in Figure 11A.2 are produced by the turbine. The definition of torque in Equation 11A.1 correctly describes the torque T_m that causes the rotation of the turbine and the synchronous generator connected to it by a shaft.

    In a rotational system, the angular acceleration due to a net torque acting on it is determined by the moment of inertia J_m of the entire system. The net torque

T_a acting on the rotating body of inertia J causes it to accelerate. Similar to systems with linear motion where $f_a = M a$, Newton's law in rotational systems becomes

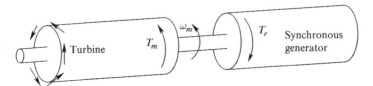

FIGURE 11A.2   Forces and torques in a turbine-generator system.

$$T_a = J_m \alpha_m \tag{11A.2}$$

where the angular acceleration $\alpha_m (= d\omega_m / dt)$ in rad/s^2 is

$$\alpha_m = \frac{d\omega_m}{dt} = \frac{T_m}{J_m} \tag{11A.3}$$

and the damping is neglected. In MKS units, a torque of 1 Nm, acting on an inertia of 1 kg·m^2, results in an angular acceleration of 1 rad/s^2.

In systems like the one shown in Figure 11A.3a, the turbine produces an electromagnetic torque T_m. The bearing friction and wind resistance (drag) can be combined with the synchronous generator torque T_e opposing the rotation. The net torque – the difference between the mechanical torque developed by the turbine and the generator torque opposing it – causes the combined inertias of the turbine and the generator to accelerate in accordance with Equation 11A.3

$$\frac{d}{dt}\omega_m = \frac{T_a}{J_m} \tag{11A.4}$$

where the net torque $T_a = T_m - T_e$ is as shown in Figure 11A.3b, and J_m is the equivalent combined inertia.

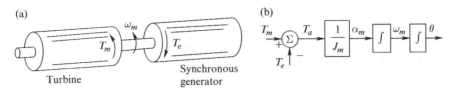

FIGURE 11A.3   Accelerating torque and acceleration. (a) Example turbine system; (b) block diagram.

Equation 11A.4 shows that the net torque is the quantity that causes acceleration, leading to changes in speed and position. Integrating the acceleration $\alpha(t)$ with respect to time,

$$\text{speed } \omega_m(t) = \omega_m(0) + \int_0^t \alpha(\tau)d\tau \tag{11A.5}$$

where $\omega_m(0)$ is the speed at $t = 0$ and τ is a variable of integration. Further, integrating $\omega_m(t)$ in Equation 11A.5 with respect to time yields

$$\theta(t) = \theta(0) + \int_0^t \omega_m(\tau)d\tau \tag{11A.6}$$

where $\theta(0)$ is the position at $t = 0$ and τ is again a variable of integration. Equations 11A.4–11A.6 indicate that torque is the fundamental variable for controlling speed and position. Equations 11A.4–11A.6 are represented in a block-diagram form in Figure 11A.3b.

In the rotational system shown in Figure 11A.4, if a net torque T_a causes the cylinder to rotate by a differential angle $d\theta$, the differential work done is

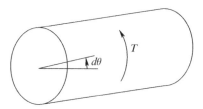

FIGURE 11A.4 Torque, work, and power.

$$dW = Td\theta \tag{11A.7}$$

If this differential rotation takes place in a differential time dt, the power can be expressed as

$$p = \frac{dW}{dt} = T\frac{d\theta}{dt} = T\omega_m \tag{11A.8}$$

where $\omega_m = d\theta/dt$ is the angular speed of rotation.

12

CONTROL OF INTERCONNECTED POWER SYSTEMS AND ECONOMIC DISPATCH

12.1 CONTROL OBJECTIVES

In an interconnected power system such as that in North America, thousands of generators, connected through hundreds of thousands of miles of transmission and subtransmission lines, operate synchronously to supply load that is constantly changing. The main advantage of a highly interconnected system is the continuity of service to consumers, ensuring reliability in case of contingencies such as unscheduled outages. An interconnected system also provides economy of operation by using optimum generation, making use of the lowest-cost generation. As we will see shortly, frequency deviations are also smaller in a highly interconnected system.

In meeting constantly changing load demand and topological changes due to outages, the network voltage and frequency are maintained by the following means:

1. Voltage regulation by excitation control of generators to control the reactive power supplied by them
2. Frequency control and maintaining the interchange of power at their scheduled values

Electric Power Systems with Renewables: Simulations Using PSS®E, Second Edition. Ned Mohan and Swaroop Guggilam.
© 2023 John Wiley & Sons, Inc. Published 2023 by John Wiley & Sons, Inc.
Companion Website: www.wiley.com/go/mohaneps

3. Optimal power flow such that the power to the load is provided in the most economical manner, considering constraints such as the transmission-line capacities and power system stability

In addition to these controls, there are supplementary controls that reduce the integral of the frequency error periodically to zero so that clocks and other appliances that depend on the grid frequency return to their normal values, and the integral of the power-exchange error is also reduced to zero.

Although the voltage and the frequency controls are implemented simultaneously in time, they act fairly independently of each other, as described next.

12.2 VOLTAGE CONTROL BY CONTROLLING EXCITATION AND REACTIVE POWER

In a power system, the quality of power is defined by keeping the voltage level at its nominal value within a fairly narrow range of ±5%, for example. This is because changes in the voltage level can disturb the load: at lower voltages, dimming lights and slowing induction motors; and at higher voltages, causing magnetic saturation of transformers and motors. As discussed in Chapter 10, a lack of reactive power can cause voltage instability and possibly voltage collapse; to avoid this, a reactive power reserve must be maintained [1].

Although there are other approaches, the primary means of voltage control is excitation control of synchronous generators in power plants. As discussed earlier, overexciting these generators allows them to supply reactive power to the system, and underexciting them lets them absorb reactive power. Other means of reactive power control to maintain voltages are the following, as discussed in Chapter 10: shunt capacitors, shunt inductive reactors, static var controllers (SVCs), static synchronous compensators (STATCOMs), series capacitors (including thyristor-controlled series capacitors), and HVDC terminals. Voltages can also be controlled by transformer tap-changing. The last-resort scheme includes automatic load shedding. However, as mentioned earlier, the primary means to obtain voltage control is automatic voltage regulation (AVR) of synchronous generators, as discussed in the following subsection.

12.2.1 Automatic Voltage Regulation (AVR) through Excitation Control

As discussed in Chapter 9, the field winding on the rotor is supplied by a DC current to establish the field flux. This requires an excitation system. There are many types of excitation systems in use, but one of the faster-acting systems is shown in Figure 12.1.

It consists of a phase-controlled thyristor rectifier supplied by three-phase AC derived from the generator output or an auxiliary supply. The DC output of the thyristor rectifier is supplied to the field winding, which rotates with the rotor, through brushes and slip rings. A voltage exciter such as the one shown in

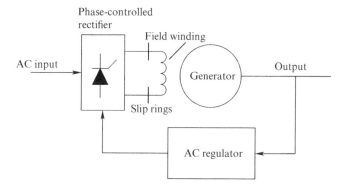

FIGURE 12.1   Field exciter for automatic voltage regulation (AVR).

Figure 12.1 can be designed to regulate the voltage at other than the generator output: for example, at a high-voltage bus after the step-up transformer. This can be accomplished with a load-compensation network that takes into account the voltage drop across the impedance between the generator and the point of regulation, primarily the transformer-leakage inductance. In such exciters, several safety features are built in to, for example, limit underexcitation to prevent exceeding the steady-state stability limit and overexcitation to prevent thermal overloading.

As discussed in Chapter 11, a fast excitation response can play a beneficial role in improving transient stability. But a power system stabilizer (PSS) should also be used to introduce damping to prevent rotor oscillations by using the rotor-speed oscillations as a signal in controlling the field excitation of the generator to maintain dynamic stability.

12.3   AUTOMATIC GENERATION CONTROL (AGC)

As mentioned earlier, the load demand fluctuates continuously and randomly, and the real power being generated must be adjusted to meet instantaneous changes in load demand. For this purpose, for example, the interconnected power system in the North American power grid is divided into four interconnections [2]. To operate safely and reliably, each interconnection comprises several control areas that continuously monitor and control the power flow. Each control area consists of either a single company or a group of companies operating many generators. These control areas are interconnected through transmission lines called *tie lines*. There is a scheduled interchange of power between the control areas to realize the benefits of having them interconnected. Most of the generators from all the areas participate in dynamically meeting changes in load demand in any of the areas. However, in a steady state, each control area handles all the changes in load demand within its own area if possible. This requires automatic generation control (AGC), which most generators are equipped with. AGC also requires a certain amount of spinning reserve that can quickly meet instantaneous changes in load demand that fluctuate randomly.

12.3.1 Load-Frequency Control

The turbines of most generators above a certain rating are governed to provide AGC. To understand turbine-governor control, consider the single turbine shown in Figure 12.2a. In such a system, if the electrical load increases, the rotor slows, resulting in reduced speed and hence reduced frequency. The frequency decrease is sensed as a feedback signal to the regulator that acts (with a negative sign) on the turbine governor to change the valve position and let in more steam. In a steady state, neglecting losses, the mechanical input, the electrical output, and the load power are all the same: $P_m = P_e = P_{Load}$.

(a)

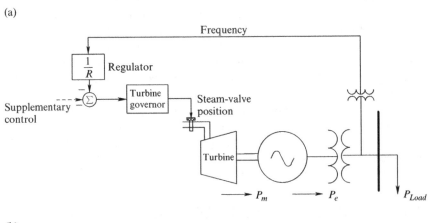

(b)

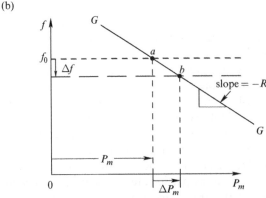

FIGURE 12.2 Load-frequency control (ignore the supplementary control at present). (a) Single turbine model; (b) load-frequency curve.

By design, the load-frequency plot is a drooping straight line GG, as shown in Figure 12.2b. Initially, at a load equal to P_m, the operating frequency is f_0. If the load increases in a steady state by an amount ΔP_m, then the frequency decreases by a value equal to Δf, which lets in more steam, as shown in Figure 12.2a. (Note that when the frequency decreases, Δf is a negative value and the regulator output is multiplied by a negative sign, thus resulting in a change in power ΔP_m that is positive to satisfy the increase in load.)

As shown in Figure 12.2b, the slope of the load-frequency control characteristic is $(-R)$, where R is called the *speed regulation* and has a positive value:

$$R(\text{in \%}) = -\frac{\Delta f\,(\text{in \%})}{\Delta P_m\,(\text{in pu})} \tag{12.1}$$

or

$$\Delta f\,(\text{in \%}) = -R(\text{in \%}) \times \Delta P_m\,(\text{in pu}) \tag{12.2}$$

For example, regulation R equal to 5% implies that a 0.1 per-unit increase in the electrical load corresponds to a 0.5% decrease in the base frequency. If the base frequency is 60 Hz, it corresponds to a decrease of 0.3 Hz: that is, the new frequency will be 59.7 Hz.

Now consider the case of two generators interconnected through a tie line, supplying the load shown in Figure 12.3a; in a steady state, neglecting losses, $Pe_1 = P_{m1}$, $P_{e2} = P_{m2}$, and $P_{m1} + P_{m2} = P_{Load}$.

(a)   (b)

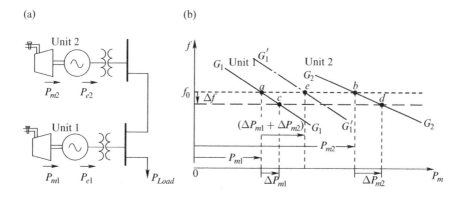

FIGURE 12.3 Responses of two generators to load-frequency control. (a) Interconnected system with a tie line; (b) load-frequency curves.

Both turbines have governor feedback mechanisms that give their units drooping characteristics, as shown by the solid lines G_1G_1 and G_2G_2 in Figure 12.3b, with regulation values R_1 and R_2. Initially, at a rated frequency f_0 of 60 Hz (or 50 Hz), the operating points are a and b, and $P_{m1} + P_{m2} = P_{Load}$. For the given characteristics G_1G_1 and G_2G_2, if the electrical load increases by an amount ΔP_{Load}, the new steady-state frequency (common to both the units) drops; from Equation 12.1,

$$\Delta f = -R_1 \Delta P_{m1} \quad \text{and} \quad \Delta f = -R_2 \Delta P_{m2} \tag{12.3}$$

and the operating points shift to c and d such that

$$\Delta P_{m1} + \Delta P_{m2} = \Delta P_{Load} \tag{12.4}$$

From Equations 12.3 and 12.4,

$$\Delta f = -\frac{\Delta P_{Load}}{(1/R_1 + 1/R_2)} \tag{12.5}$$

In general, if many such generators are interconnected, from Equation 12.5,

$$\Delta f = -\frac{1}{\sum_i 1/R_i} \Delta P_{Load} \tag{12.6}$$

Therefore, comparing Equation 12.6 with Equation 12.2, the equivalent speed regulation is

$$R_{eq} = -\frac{1}{\sum_i 1/R_i} \tag{12.7}$$

and the change in frequency for a given change in load is much smaller in an interconnected system compared to a single-generator case. Consequently, interconnected generators result in a much "stiffer" system where all the generators initially participate in accommodating the change in load, and thus the change in frequency is very small.

Example 12.1
Consider two generators in parallel, operating at 60 Hz and having widely different regulation, with $R_1 = 5\%$ and $R_2 = 16.7\%$. A load change of 0.1 pu occurs. Calculate the equivalent value of the regulation, the initial decrease in frequency, and how the change in load is initially shared by the two generators.

Solution   From Equation 12.7, R_{eq} = 3.85 (in %). Therefore, from Equation 12.6,

$$\Delta f = -0.385 \text{ (in %)} = -0.231 \text{ Hz}$$

After the load change, the new frequency is initially 59.77 Hz. Using Equation 12.1,

$$\Delta P_{m1} = \frac{0.385}{5.0} = 0.077 \text{ pu} \quad \text{and} \quad \Delta P_{m2} = \frac{0.385}{16.7} = 0.023 \text{ pu}$$

which shows that unit 1, with a smaller value of regulation, initially picks up a greater share of the load.

12.3.2   Automatic Generation Control (AGC) and Area Control Error (ACE)

Earlier, we looked at the initial response of interconnected generator units to a change in load. Now consider two control areas interconnected through a tie line, as shown in Figure 12.4, where each area may consist of several generator units. In each control area, for discussion purposes, all the generators are combined as an equivalent single generator unit, similar to that in Figure 12.3a.

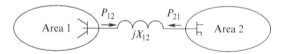

FIGURE 12.4   Two control areas.

To restore the frequency and tie-line flow to their original and scheduled values in the system in Figure 12.3a, where the load change occurs on unit 1, an AGC supplementary control raises the generation characteristic of unit 1 in a steady state to $G_1'G_1'$, as shown in Figure 12.3b. The entire load change is thus provided by unit 1, and the operating point for unit 1 shifts to a new point e, whereas the operating point for unit 2 returns to its original value, b.

To do this, for each area, an *area control error* (ACE) is defined as the sum of the tie-line flow deviation and the frequency deviation multiplied by a frequency-bias factor B. Therefore, defining the increase in power flow out of one area to the other area – for example, the increase ΔP_{12} from area 1 to area 2 – to be positive,

$$ACE_1 = \Delta P_{12} + B_1\Delta f \tag{12.8}$$

Similarly,

$$ACE_2 = \Delta P_{21} + B_2 \Delta f \quad \text{(where, neglecting losses, } \Delta P_{21} = -\Delta P_{12}) \quad (12.9)$$

A negative value of ACE for an area indicates that there is not enough generation in that area. In the block diagram in Figure 12.4, following a load change in either of the two areas, the final steady state is reached only if both ACE values, as defined by Equations 12.8 and 12.9, come to zero when the tie-line flow and the frequency are restored to their original values. This implies that in a steady state, an area with a load change completely absorbs its own load change. This was illustrated earlier, in Figure 12.3b, where the change in load on unit 1 results in shifting the load-frequency characteristic to $G_1'G_1'$ such that it supplies the entire change in that area, and the unit 2 generation remains unchanged at $G_2\,G_2$.

Each generator unit participating in AGC has a supplementary controller with the ACE as the input, as shown in Figure 12.5. The supplementary controller output and regulation feedback act on the governor to change the steam-valve position, as shown in Figure 12.5.

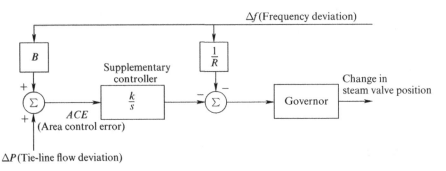

FIGURE 12.5   Area control error (ACE) for automatic generation control (AGC).

In a steady state, the results are the same regardless of the values of the frequency-bias settings B_1 and B_2 in Equations 12.8 and 12.9, provided the control is stable. During dynamic operation, field experience has shown that selecting a frequency-bias factor of approximately $B = 1/R$ in each area provides good dynamic results. When a control area has more than one tie line, as is almost always the case, the AGC action described earlier brings the net control-area exchange to its original scheduled value. This is illustrated in the example power system in Figure 12.6, where two control areas are connected through two tie lines. There are two revenue meters (M) to measure power flows at the boundaries of the two control areas. Therefore, in defining ACE in Equations 12.8 and 12.9, the tie-line deviation between the two control areas is $\Delta P_{12} = \Delta P_{1\text{-}2} + \Delta P_{1\text{-}3}$, where $\Delta P_{1\text{-}2}$ and $\Delta P_{1\text{-}3}$ are the line flows at the boundaries of the two control areas.

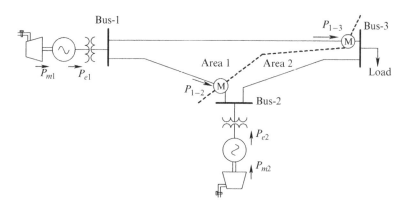

FIGURE 12.6 Two control areas in the example power system with three buses.

Example 12.2
Consider the example three-bus power system described in Chapter 5. Ignoring the line losses, calculate the power flow on the three lines. Repeat this if the load increases to 600 MW (6 pu) but, due to AGC applied to both units, the net power flow between the two areas remains the same.

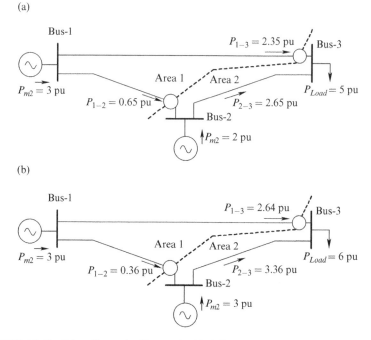

FIGURE 12.7 Line flows in Example 12.2 for (a) a 500 MW load and (b) a 600 MW load.

Solution Using the Python program developed in Chapter 5 and reducing all line resistances to zero, the line flows are shown in Figure 12.7a for a 500 MW load. In the case of 600 MW with both units under AGC, the net flow from area 1 to area 2 remains the same, and the entire load change in area 2 is absorbed by unit 2. The line flows are shown in Figure 12.7b.

In practice, the AGC is run every 2 to 4 seconds, and the ACE must be brought to zero within every 10-minute interval.

12.3.3 Dynamic Performance of Interconnected Areas

So far, we have examined the interconnections of various areas on a steady-state basis. However, the load can change as a step, and the interconnected system reacts dynamically. In doing so, it is important that there is enough damping; otherwise, the system can become unstable. The dynamic response of the system depends on many parameters: inertias of the interconnected systems, damping, regulators, gain values of the supplementary controller in Figure 12.5 to correct the ACE, etc.

Another important parameter is the magnitude of the synchronizing torque coefficient between the areas. This can be explained using the simple two-area example shown in Figure 12.8, where all the generating units in a control area are represented by an equivalent generator since the objective is to study the inter-area oscillations, not the intra-area oscillations.

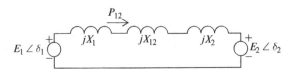

FIGURE 12.8 Electrical equivalent of a two-area interconnection.

In Figure 12.8, X_1 and X_2 are the reactances of the equivalent generators, with E_1 and E_2 as the internal emfs and X_{12} as the tie-line reactance. Thus,

$$X_T = X_1 + X_{12} + X_2 \tag{12.10}$$

The power flow from area 1 to area 2 is

$$P_{12} = \frac{E_1 E_2}{X_T} \sin \delta_{12} \tag{12.11}$$

where $\delta_{12} = \delta_1 - \delta_2$. Therefore, linearizing Equation 12.11 around a steady-state operating point with initial power P_0 and initial angle difference δ_0, we can write Equation 12.11 as

$$P_0 + \Delta P_{12} = \frac{E_1 E_2}{X_T} \sin(\delta_0 + \Delta \delta_{12}) \tag{12.12a}$$

Noting that $\sin(a + b) = \sin a \cdot \cos b + \cos a \cdot \sin b$, expanding the sine term in Equation 12.12a results in the following for a small value of the perturbation $\Delta \delta_{12}$:

$$P_0 + \Delta P_{12} = \frac{E_1 E_2}{X_T} \left(\sin \delta_0 \cdot \underbrace{\cos \Delta \delta_{12}}_{(\approx 1)} + \cos \delta_0 \cdot \underbrace{\sin \Delta \delta_{12}}_{(\approx \Delta \delta_{12})} \right) \tag{12.12b}$$

Therefore, from Equation 12.12b, recognizing that $\Delta \delta_{12} = \Delta \delta_1 - \Delta \delta_2$,

$$\Delta P_{12} = \underbrace{\left(\frac{E_1 E_2}{X_T} \cos \delta_0 \right)}_{T_{12}} (\Delta \delta_1 - \Delta \delta_2) \tag{12.13}$$

where if the speed is expressed in per-unit as being equal to unity, the quantity within the brackets is the synchronizing torque coefficient T_{12} between areas 1 and 2:

$$T_{12} = \frac{E_1 E_2}{X_T} \cos \delta_0 \tag{12.14}$$

Equation 12.14 shows that the smaller the initial operating angle δ_0, the larger the magnitude of the synchronizing torque coefficient that determines the period and the magnitude of the tie-line oscillations following a load change. These inter-area oscillations in a two-area system are illustrated by the following example.

Example 12.3
Consider the two-area system shown in Figure 12.9, where both areas are identical and under AGC. The system parameters are specified in a MATLAB file associated with this example on the accompanying website. There is a step change of load in area 1. Plot the following: ΔP_{m1}, ΔP_{m2}, and ΔP_{12}.

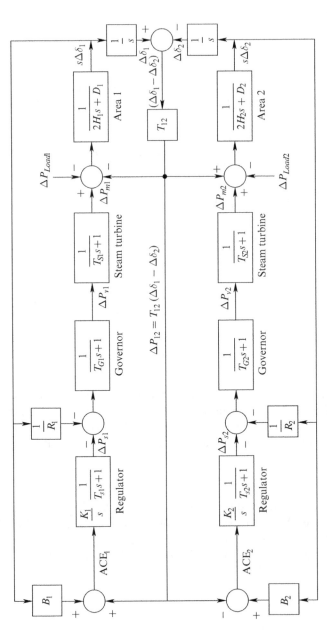

FIGURE 12.9 Two-area system with AGC. Source: Adapted from [3].

Solution This system is modeled in Simulink and included on the accompanying website. The results, plotted in Figure 12.10, show that for a load change in area 1, both areas participate initially; but in a steady state, the load change is entirely satisfied by area 1, and hence $\Delta P_{m1} = \Delta P_{\text{Load1}}$ and $\Delta P_{m2} = 0$. The tie-line power deviation ΔP_{12} oscillates and eventually settles down to zero. The frequency of these oscillations depends on the system parameters.

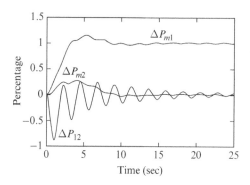

FIGURE 12.10 Simulink results of the two-area system with AGC in Example 12.3.

12.4 ECONOMIC DISPATCH AND OPTIMUM POWER FLOW

One of the advantages of an interconnected system is the optimum allocation of generation for the least overall production cost, ensuring that the transmission line loadings are within their capacities and the system's transient stability margins are maintained.

12.4.1 Economic Dispatch

In power plants, the cost of generating electricity depends on fixed operating costs (which depend on capital investment, etc., and are independent of the power being produced) and variable operating costs, including fuel costs (which depend on the power being produced). However, once a power plant has been built, the operating strategy from an economic point of view is to minimize the total fuel cost to generate the required amount of power.

In the normal operating range, the efficiency of a thermal power plant increases somewhat with increasing power levels. In other words, the heat rate – the primary energy in MBTUs (million British thermal units) consumed per hour divided by the electrical power P – decreases slightly with the power level, as shown in Figure 12.11.

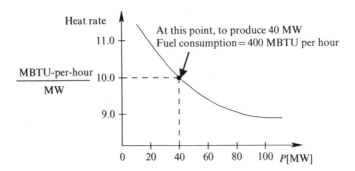

FIGURE 12.11   Heat rate at various generated power levels.

Contrary to what the heat-rate curve in Figure 12.11 suggests, in practice, due to various considerations, the fuel cost is taken to increase with power, as shown in Figure 12.12a.

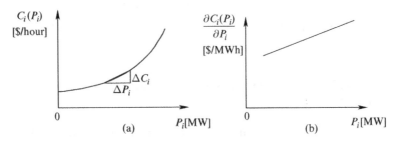

FIGURE 12.12   (a) Fuel cost and (b) marginal cost as functions of power output.

In general, the fuel-cost curve for a unit i can be expressed as a quadratic function of the power Pi that is generated:

$$C_i(P_i) = a_i + b_i P_i + c_i P_i^2 \tag{12.15}$$

The slope of the fuel-cost curve in Figure 12.12a at any power level is the marginal cost, which is the cost of generating an additional 1 MWh at a particular power level. This marginal cost, in dollars/MWh, can also be calculated by taking the partial derivative of the cost in Equation 12.15 with respect to P_i:

$$\frac{\partial C_i(P_i)}{\partial P_i} = b_i + 2c_i P_i \tag{12.16}$$

This is a straight line with an upward (positive) slope, as shown in Figure 12.12b.

Consider an example where an area has three generators. Then there are three marginal cost equations similar to Equation 12.16 and three marginal-cost curves, as plotted in Figure 12.13.

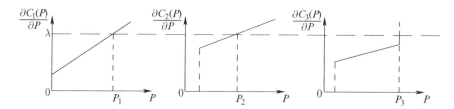

FIGURE 12.13   Marginal costs for the three generators.

Another equation is based on power balance – the sum of the generation must equal the sum of the load and the losses:

$$\sum_i P_i = P_{\text{Load}} + P_{\text{Losses}} \qquad (12.17)$$

We want to calculate the values of P_1, P_2, and P_3 that result in the total minimum fuel cost while the power balance equation is satisfied. An intuitive solution, and correctly so, is to operate such that all three generators have the same marginal cost and the power balance is satisfied. The reason is as follows: if one of the generators operates at a higher marginal cost, shifting generation from it to the lower-marginal-cost generators will result in a lower overall cost. This will continue until all the generators have equal marginal costs, as plotted in Figure 12.13.

A more formal solution to this well-known optimization problem can be obtained using what is called the *Lagrangian multiplier method*. A Lagrangian cost function is defined as follows for this three-generator system

$$L = C_1(P_1) + C_2(P_2) + C_3(P_3) - \lambda[P_1 + P_2 + P_3 - (P_{\text{Load}} + P_{\text{Losses}})] \quad (12.18)$$

where λ is the Lagrangian multiplier. This Lagrangian function has four variables: P_1, P_2, P_3, and λ, assuming P_{Losses} to be constant. To minimize this Lagrangian cost function, we take its partial derivative with respect to each of these four variables and set them to zero. Therefore, from Equation 12.18,

$$\frac{\partial L}{\partial P_1} = \frac{\partial C_1(P_1)}{\partial P_1} - \lambda = 0 \qquad (12.19)$$

$$\frac{\partial L}{\partial P_2} = \frac{\partial C_2(P_2)}{\partial P_2} - \lambda = 0 \tag{12.20}$$

$$\frac{\partial L}{\partial P_3} = \frac{\partial C_3(P_3)}{\partial P_3} - \lambda = 0 \tag{12.21}$$

$$P_1 + P_2 + P_3 - (P_{Load} + P_{Losses}) = 0 \left(\because \frac{\partial L}{\partial \lambda} = 0 \right) \tag{12.22}$$

The last equation is the power-balance equation. From Equations 12.19–12.22, the cost function in Equation 12.18 is minimized if

$$\frac{\partial C_1(P_1)}{\partial P_1} = \frac{\partial C_2(P_2)}{\partial P_2} = \frac{\partial C_3(P_3)}{\partial P_3} = \lambda \quad \text{(assuming } P_{Losses} \text{ to be constant) (12.23)}$$

The solution given by Equation 12.23 is the same as the intuitive solution: the overall cost is minimized if all three generators operate at equal marginal costs while satisfying the power balance. Therefore, from equations similar to Equation 12.16,

$$b_1 + 2c_1 P_1 = \lambda \tag{12.24}$$

$$b_2 + 2c_2 P_2 = \lambda \tag{12.25}$$

$$b_3 + 2c_3 P_3 = \lambda \tag{12.26}$$

and

$$P_1 + P_2 + P_3 = P_{Load} + P_{Losses} \tag{12.27}$$

From these four equations, the four variables P_1, P_2, P_3, and λ can be solved, where λ is the system's optimum marginal cost in \$/MWh (that is, the cost in \$/hour of supplying an increase of 1 MW in system load and losses). In practice, there is a maximum limit on the power that can be produced by a generator and also a lower limit that a unit must produce unless taken offline. These limits can be indicated by vertical lines in the marginal cost curves, as shown in Figure 12.13.

Example 12.4
In a control area, there are two generators, 100 MW each. The marginal costs for these two generators can be as expressed as follows:

$$\frac{\partial C_1}{\partial P_1} = 1.8 + 0.01 \, P_1 \quad \text{(in \$/MWh)}$$

$$\frac{\partial C_2}{\partial P_2} = 1.5 + 0.02 \, P_2 \quad \text{(in \$/MWh)}$$

If this area has to supply a total of 150 MW, calculate the area's optimum marginal cost λ and the power supplied by each generator.

Solution Equating the two marginal costs and equating the sum of the two powers to 150 MW,

$$1.8 + 0.01\, P_1 = 1.5 + 0.02\, P_2 \quad \text{and} \quad P_1 + P_2 = 150$$

Solving these two equations, P_1 = 90 MW and P_2 = 60 MW. From any of Equations 12.24–12.26, the marginal cost is λ = 2.7 ($/MWh).

12.4.2 Unit Commitment and Spinning Reserve

The economic dispatch just discussed is based on the generation capacity put into service. However, other factors dictate the capacity that must be in service for a given load for angle and voltage stability and to quickly meet load changes in a control area (now referred to as a *balancing authority*). Each control area must maintain a certain level – 15 to 20%, for example – of *spinning reserve*, which represents the aggregate power plant capacity in a control area that is not utilized at a given time and can respond immediately being online, because a cold start, especially of thermal plants, may take tens of minutes or even hours. If this spinning reserve is insufficient, a new generator unit may be committed, called the *unit commitment* [4–6].

12.4.3 Optimal Power Dispatch and Flow

In the earlier analysis of economic dispatch to minimize overall fuel cost, the transmission line losses are represented as constant, independent of the values of P_1, P_2, and P_3. In practice, this is not the case, since a distant generator will incur higher transmission line losses, and the cost of these losses must be included in the cost-minimization problem. Also, no consideration was given to transmission line capacities, bus voltage limits, and the system's transient stability. When constraints such as these are factored in, the cost minimization is called *optimal power dispatch*; under those conditions, power flow on various lines is the *optimal power flow* (OPF) [5].

PSS®E offers a wide range of options for objective functions, controls, and settings for optimal power flow. It can perform economic dispatch by taking into consideration various generator costs. We can specify the bus voltage constraints; different types of cost curves such as piece-wise linear, piece-wise quadratic, polynomial, and exponential costs; branch flow constraints; inter-area flow constraints; and many more. It can also perform unit commitment operations. More details on the OPF and homework problems can be found on the accompanying website.

REFERENCES

1. P. Kundur. 1994. *Power System Stability and Control*. McGraw Hill.
2. U.S. Department of Energy. https://www.energy.gov/oe/services/electricity-policy-coordination-and-implementation/transmission-planning/recovery-act-0.
3. L. K. Kirchmayer. 1959. *Economic Control of Interconnected Systems*. John Wiley & Sons.
4. L. K. Kirchmayer. 1958. *Economic Operation of Power Systems*. John Wiley & Sons.
5. A. J. Wood, B. F. Wollenberg, and G. B. Sheblé. 2013. *Power generation, operation, and control*. John Wiley & Sons.
6. N. Cohn. 1967. *Control of Generation and Power Flow on Interconnected Systems*. John Wiley & Sons.

PROBLEMS

12.1 Three generators operating at 60 Hz are connected in parallel and have the following regulation values: $R_1 = 5\%$, $R_2 = 10\%$, and $R_3 = 15\%$. A load change of 0.1 pu occurs. Calculate the equivalent value of the regulation, the initial decrease in frequency, and how the change in load is shared by the three generators initially.

12.2 Repeat Example 12.2 using MATLAB and PSS®E to compute the line flows if the load on bus 3 decreases from 500 MW to 400 MW.

12.3 Repeat Example 12.3 using Simulink, eliminating the low-pass filtering in the regulators by setting $T_{s1} = T_{s2} = 0$. Compare the results with those in Example 12.3.

12.4 The marginal costs in \$/MWh of three generators can be expressed as follows: $(1.0 + 0.02P)$ for generator 1, $(2.0 + 0.015P)$ for generator 2, and $(1.5 + 0.01P)$ for generator 3, where P is in MW. The total power supplied by these generators is 500 MW. Calculate the system marginal cost λ and the load shared by each of the three generators.

12.5 Repeat Problem 12.4 if the power output of generator 3 is limited to a range of 25 to 75 MW.

12.6 The daily load-duration curve over a 24-hour period is shown in Figure P12.6 and varies from 200 to 600 MW. It is supplied by three generators

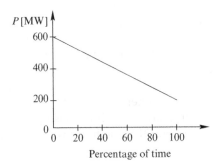

FIGURE P12.6 Load-duration curve.

whose marginal cost curves are described in Problem 12.4. If an economic dispatch is implemented, calculate the daily fuel cost in dollars for each generator and the total cost to meet this power demand.

PSS®E-BASED PROBLEMS

12.7   Using Example 5.4 and considering line-charging capacitance, perform OPF with load at 550 MW, 110 MVAR. The line thermal limits are not enforced. Use the cost parameters as defined on the accompanying website.

12.8   Using Example 5.4 and considering line-charging capacitance, perform OPF with load at 600 MW, 120 MVAR. The line thermal limits are not enforced. Use the cost parameters as defined on the accompanying website. Compare the results with those of Problem 12.7.

12.9   Using Example 5.4 and considering line-charging capacitance, perform OPF with load at 600 MW, 120 MVAR. The line thermal limits are enforced. Use the cost parameters as defined on the accompanying website. Compare the results with those of Problems 12.7 and 12.8.

13

TRANSMISSION LINE FAULTS, RELAYING, AND CIRCUIT BREAKERS

Transmission lines stretch over large distances and are subject to faults involving one or more phases and ground. Such faults can cause momentary power outages and, more importantly – if protective action is not taken, can cause permanent damage to transmission equipment such as the lines and transformers. In this chapter, we will analyze the causes and types of faults, relays to detect them, and circuit breakers to isolate them and restore the power system to normal.

13.1   CAUSES OF TRANSMISSION LINE FAULTS

A common cause of such faults is tree branches near the right of way falling on transmission lines and shorting them to ground. For example, sagging transmission lines touching trees initiated the major blackout on 14 August 2003 in the northeast United States. Another common fault occurrence is due to backflash when the transmission line tower or one of the ground wires is struck by lightning – which represents a current source of several thousand kilo-amperes. This current flowing through the tower footing impedance can raise the tower potential above the local ground to such a level that without surge arresters (discussed in the next chapter), the insulator strings may flash over.

The lightning current is momentary and lasts just a few tens of microseconds, but the arc established by such an insulator-string flashover results in a

Electric Power Systems with Renewables: Simulations Using PSS®E, Second Edition. Ned Mohan and Swaroop Guggilam.
© 2023 John Wiley & Sons, Inc. Published 2023 by John Wiley & Sons, Inc.
Companion Website: www.wiley.com/go/mohaneps

short to ground of the power-frequency voltages through the arc impedance. If short-circuit currents are not detected by a relay that sends a signal to open circuit breakers to interrupt this current, the current at the power frequency will keep flowing until the equipment is damaged – for example, by fire. Therefore, it is essential that these faults are detected and that circuit breakers interrupt them quickly. A few cycles after the fault has been cleared, the circuit breakers can be reclosed and normal system operation can resume.

We analyze short-circuit faults to (i) set the relays so they can detect such faults and (ii) make sure the circuit-breaker ratings are such that they are capable of interrupting the fault currents.

13.2 SYMMETRICAL COMPONENTS FOR FAULT ANALYSIS

More than 80% of faults involve just one of the three phases and ground: for example, a tree branch touching or falling on one of the phases. Such faults are often unsymmetrical since not all three phases are involved.

We will begin with a discussion of the analytical approach to analyzing unsymmetrical faults, of which symmetrical faults are a subset. Consider the fault location f shown in Figure 13.1. To keep this discussion simple yet practical, although the fault may be unsymmetrical, we will assume that the rest of the system seen from the fault location is balanced. Such a fault results in fault currents i_a, i_b, and i_c as shown in Figure 13.1a. During the fault period, even though the power-system voltages and currents are in their transient state, we will assume that they have reached a pseudo-steady state at their line frequency so that we can use phasors to describe them. Therefore, in such a steady state at the line frequency, i_a, i_b, and i_c in Figure 13.1a can be represented by phasors $\overline{I}_a$, $\overline{I}_b$, and $\overline{I}_c$, as shown in Figure 13.1b.

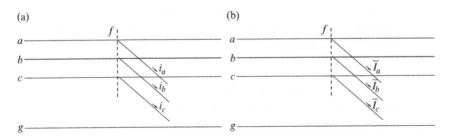

FIGURE 13.1   Fault in a power system. (a) Instantaneous currents; (b) phasor currents.

13.2.1 Calculating the Symmetrical Components

To analyze a power system with an unsymmetrical fault, Fortesque [1] showed long ago that unbalanced currents can be expressed as sums of components that are symmetrical and balanced. Assuming the power system to be a linear network where the principle of superposition can be applied, the voltages and currents in the faulted system can be obtained by summing the components in each of the balanced sequence networks.

As an example, the unbalanced fault currents $\bar{I}_a$, $\bar{I}_b$, and $\bar{I}_c$ are shown in Figure 13.2. They can be shown to be composed of three components

$$\begin{aligned}
\bar{I}_a &= \bar{I}_{a1} + \bar{I}_{a2} + \bar{I}_{a0} \\
\bar{I}_b &= \bar{I}_{b1} + \bar{I}_{b2} + \bar{I}_{b0} \\
\bar{I}_c &= \bar{I}_{c1} + \bar{I}_{c2} + \bar{I}_{c0}
\end{aligned} \tag{13.1}$$

where the subscript 1 refers to the positive-sequence, 2 to the negative-sequence, and 0 to the zero-sequence components.

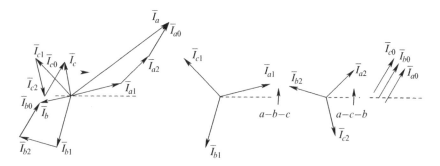

FIGURE 13.2   Sequence components.

Within each sequence, the three phases are balanced and sinusoidal at the line frequency. Positive-sequence components $\bar{I}_{a1}$, etc., are balanced in amplitude and are in the same a-b-c sequence as the power system voltages and currents: that is, phase $\bar{I}_{b1}$ lags $\bar{I}_{a1}$ by 120 degrees, and so on. The negative-sequence components are also balanced in amplitude but are in the sequence a-c-b: that is, $\bar{I}_{c2}$ lags $\bar{I}_{a2}$ by 120 degrees, and so on. The zero-sequence components are also balanced in amplitude, but in contrast to both the positive- and the negative-sequence components, $\bar{I}_{a0}$, $\bar{I}_{b0}$, and $\bar{I}_{c0}$ have the same phase and thus are equal to each other – hence the name *zero-sequence*. Given these properties of the sequence components, for the ease of writing them, we will define the following operators:

$$\begin{aligned}
a &= 1\angle 120° = -0.5 + j0.866 \\
a^2 &= 1\angle 240° = -0.5 - j0.866
\end{aligned} \tag{13.2}$$

Therefore, the sequence components in phases b and c in Equation 13.1 can be written in terms of the phase a components as follows:

$$\overline{I}_{b1} = a^2 \overline{I}_{a1}; \ \overline{I}_{c1} = a\overline{I}_{a1}$$
$$\overline{I}_{b2} = a\overline{I}_{a2}; \ \overline{I}_{c2} = a^2 \overline{I}_{a2}$$

(13.3)

Substituting, the fault currents in terms of phase a components can be written as

$$\overline{I}_a = \overline{I}_{a1} + \overline{I}_{a2} + \overline{I}_{a0}$$
$$\overline{I}_b = a^2 \overline{I}_{a1} + a\overline{I}_{a2} + \overline{I}_{a0}$$
$$\overline{I}_c = a\overline{I}_{a1} + a^2 \overline{I}_{a2} + \overline{I}_{a0}$$

(13.4)

which can be written in matrix form as

$$\begin{bmatrix} \overline{I}_a \\ \overline{I}_b \\ \overline{I}_c \end{bmatrix} = \begin{bmatrix} 1 & 1 & 1 \\ a^2 & a & 1 \\ a & a^2 & 1 \end{bmatrix} \begin{bmatrix} \overline{I}_{a1} \\ \overline{I}_{a2} \\ \overline{I}_{a0} \end{bmatrix}$$

(13.5)

Inverting the matrix in Equation 13.5, we can solve for the three sequence components of the phase a current in terms of the fault currents:

$$\begin{bmatrix} \overline{I}_{a1} \\ \overline{I}_{a2} \\ \overline{I}_{a0} \end{bmatrix} = \frac{1}{3} \begin{bmatrix} 1 & a & a^2 \\ 1 & a^2 & a \\ 1 & 1 & 1 \end{bmatrix} \begin{bmatrix} \overline{I}_a \\ \overline{I}_b \\ \overline{I}_c \end{bmatrix}$$

(13.6)

Example 13.1
In Figure 13.2, the three-phase the following values in per-unit: $\overline{I}_a = 2.2\angle 26.6°\,\text{A}$, $\overline{I}_b = 0.6\angle -156.8°\,\text{A}$, $\overline{I}_c = 0.47\angle 138.7°\,\text{A}$. Calculate the symmetrical components $\overline{I}_{a1}, \overline{I}_{a2}$, and $\overline{I}_{a0}$.

Solution Applying Equation 13.6 with the given values of the phase currents, these phase currents are made of symmetrical components with the positive sequence $\overline{I}_{a1} = 1.0\angle 15°\text{A}$, negative sequence $\overline{I}_{a2} = 0.75\angle 30°\text{A}$, and zero sequence $\overline{I}_{a0} = 0.5\angle 45°\text{A}$. MATLAB and Python solution is provided on the accompanying website.

13.2.2 Applying the Sequence Components to the Network and Superposition

Looking into the system from the fault point for each sequence, the system is balanced. Therefore, the three-phase system can be represented on a per-phase basis, as discussed in Chapter 3: for example, in terms of phase a. For each sequence, the power-system network can be drawn on a per-phase basis, as shown in Figure 13.3. The connection of these sequence networks depends on the type of fault. Once the currents and voltages are calculated in these sequence networks for phase a, they can be used to calculate voltages and currents in the faulted system using equations similar to Equation 13.5.

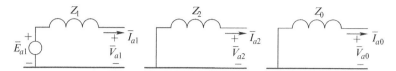

FIGURE 13.3 Sequence networks.

Note that only the positive-sequence network in Figure 13.3 has an internal emf that equals the Thevenin voltage seen from the fault point, looking into the network. The other sequence networks do not have internal emfs because of the assumption of a balanced network before the fault.

13.3 TYPES OF FAULTS

There can be many different types of faults, and the procedure for solving for all of them is similar. In this chapter, the following faults are considered:

- Symmetrical three-phase, and three-phase-to-ground fault
- Single-line-to-ground fault
- Double-line-to-ground fault
- Double-line fault (ground is not involved)
- Fault with fault impedances

In addition to these short-circuit faults, there can be instances of open-circuit conductor(s) that can be serious safety issues. Analysis of these is left as a homework exercise.

13.3.1 Symmetrical Three-Phase and Three-Phase-to-Ground Faults

In a balanced system, three-phase and three-phase-to-ground faults are identical. In Figure 13.4a, the voltage at the faulted point is zero with respect to ground, and the current flowing into ground is zero since $\bar{I}_a + \bar{I}_b + \bar{I}_c = 0$.

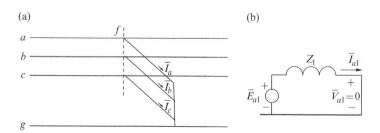

FIGURE 13.4 Three-phase symmetrical fault. (a) Faulted system; (b) equivalent circuit.

Since these three fault currents form a balanced three-phase set, their negative-and zero-sequence components are zero, and only the positive-sequence network on a per-phase basis needs to be considered; this is shown in Figure 13.4b, where $\overline{I}_a = \overline{I}_{a1}$ since the other two components are zero.

13.3.2 Single-Line-to-Ground Fault

This is shown in Figure 13.5a, where phase a is faulted to ground through a fault impedance Z_f. For this fault,

$$\overline{I}_b = \overline{I}_c = 0 \tag{13.7}$$

$$\overline{V}_a = Z_f \overline{I}_a \tag{13.8}$$

Substituting $\overline{I}_b$ and $\overline{I}_c$ as zero in (13.6),

$$\overline{I}_{a1} = \overline{I}_{a2} = \overline{I}_{a0} \tag{13.9}$$

And thus, from 13.4,

$$\overline{I}_{a1} = \frac{\overline{I}_a}{3} \tag{13.10}$$

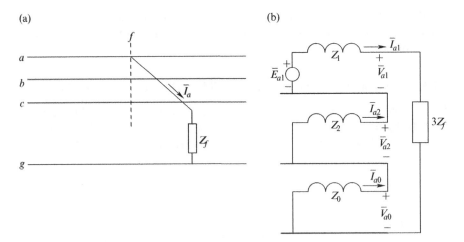

FIGURE 13.5 Single-line-to-ground fault.

From Equation 13.8, in terms of symmetrical components,

$$\overline{V}_a = \overline{V}_{a1} + \overline{V}_{a2} + \overline{V}_{a0} = Z_f \overline{I}_a \tag{13.11}$$

Substituting for $\bar{I}_a$ from Equation 13.10 into Equation 13.11,

$$\bar{V}_a = 3Z_f \bar{I}_{a1} \tag{13.12}$$

Equations 13.9 and 13.12 are satisfied by connecting the three sequence networks in series, as shown in Figure 13.5b, from which

$$\bar{I}_{a1} = \bar{I}_{a2} = \bar{I}_{a0} = \frac{\bar{E}_{a1}}{Z_1 + Z_2 + Z_0 + 3Z_f} \tag{13.13}$$

Knowing the sequence currents, all three sequence networks can be solved, and any of the currents and voltages can be computed in the faulted network.

13.3.3 Double-line-to-Ground Fault

This is shown in Figure 13.6a, where phases b and c are faulted to ground. The fault conditions result in the following conditions:

$$\bar{I}_a = 0 \tag{13.14}$$

$$\bar{V}_b = \bar{V}_c = 0 \tag{13.15}$$

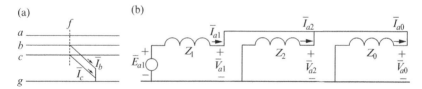

FIGURE 13.6 Double line-to-ground fault. (a) Faulted system; (b) equivalent circuit.

Using the condition in Equation 13.15 for voltages in an equation for voltages similar to Equation 13.6,

$$\bar{V}_{a1} = \bar{V}_{a2} = \bar{V}_{ao} \tag{13.16}$$

From Equation 13.16, it is clear all three sequence networks are in parallel, as shown in Figure 13.6b. Applying Kirchhoff's current law in this circuit, Equation 13.14 is satisfied in light of Equation 13.4: that is, $\bar{I}_a = \bar{I}_{a1} + \bar{I}_{a2} + \bar{I}_{a0} = 0$.

13.3.4 Double-Line Fault (Ground Is Not Involved)

This is shown in Figure 13.7a, where phases b and c are shorted to each other through Z_f. This fault results in the following conditions:

$$\bar{I}_a = 0 \qquad (13.17)$$

$$\bar{I}_b = -\bar{I}_c \qquad (13.18)$$

$$\bar{V}_b = \bar{V}_c + Z_f \bar{I}_b \qquad (13.19)$$

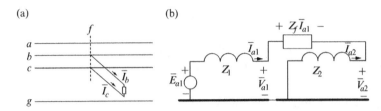

FIGURE 13.7 Double-line fault (ground not involved).

By observation of the faulted system, there is no connection to ground, so there cannot be any zero-sequence current

$$\bar{I}_{a0} = 0 \qquad (13.20)$$

and hence, in the zero-sequence network in Figure 13.3,

$$\bar{V}_{a0} = 0 \qquad (13.21)$$

Since $\bar{I}_a$ and $\bar{I}_{a0}$ are both zero, from Equation 13.4,

$$\bar{I}_{a1} = -\bar{I}_{a2} \qquad (13.22)$$

Therefore, from Equation 13.5,

$$\bar{I}_b = (a^2 - a)\bar{I}_{a1} \qquad (13.23)$$

From the voltage fault condition of Equation 13.19, with $\bar{V}_{a0} = 0$, from the voltage equation similar to Equation 13.5 for $\bar{V}_b$, using Equation 13.23,

$$\underbrace{a^2\bar{V}_{a1} + a\bar{V}_{a2}}_{(=\bar{V}_b)} = \underbrace{a\bar{V}_{a1} + a^2\bar{V}_{a2}}_{(=\bar{V}_c)} + Z_f(a^2 - a)\bar{I}_{a1} \qquad (13.24)$$

From this equation,

$$\bar{V}_{a1} = \bar{V}_{a2} + Z_f\bar{I}_{a1} \qquad (13.25)$$

Using Equations 13.18 and 13.25, the sequence networks are connected as shown in Figure 13.7b.

13.4 SYSTEM IMPEDANCES FOR FAULT CALCULATIONS

To calculate currents under unsymmetrical faults, power-system components must be represented by their appropriate impedances for all three sequences: positive, negative, and zero.

13.4.1 Transmission Lines

Transmission lines are assumed to be perfectly transposed, and their positive-sequence and negative-sequence impedances are the same. Their zero-sequence impedance involving ground return is greater in value and can be calculated using a line-constants program such as EMTDC, as described in Chapter 14.

13.4.2 Simplified Synchronous Generator Representation

To calculate fault currents within a few cycles of the fault inception, the sub-transient reactances of the generators (discussed in Chapter 9) are used. Assuming that it is a round-rotor machine, the positive-sequence impedance equals X_d'' for the d-axis, whereas that of the q-axis subtransient reactance may be slightly smaller. Typically, $X_d'' = 0.12 - 0.25\,\mathrm{pu}$.

In the negative sequence, the voltage and currents are of the negative sequence a-c-b, while the rotor turns at the synchronous speed in the forward direction dictated by the positive sequence a-b-c excitation. In this situation, the armature-reaction mmf produced by the negative-sequence currents rotates in the direction opposite to the rotor. This is at twice the synchronous speed with respect to the rotor. Therefore, the d-axis and q-axis damper windings on the rotor shield the flux due to the negative-sequence armature-reaction flux from going past the damper windings. The negative-sequence reactance of the synchronous generator can be written as

$$X_2 = \frac{X_d'' + X_q''}{2} \tag{13.26}$$

Typically, $X_2 \simeq X_d''$.

The zero-sequence impedance depends on the leakage reactance per-phase plus three times the impedance, which is usually connected from the neutral to ground of generators through a transformer. If for some reason the neutral is floating with respect to ground and this neutral impedance is infinite, the zero-sequence impedance is also infinite, indicating that the zero-sequence current cannot flow through the generator. Typically, the internal zero-sequence reactance of a turbo-generator can be approximated as $X_0 = 0.5X_2$.

For fault calculations, the positive-sequence network for the generator is usually represented by the subtransient reactance and a voltage source behind it, such that together they yield the appropriate pre-fault voltages and currents in the network, as discussed in Chapter 9.

13.4.3 Transformer Representation in Fault Studies

Generally, only the leakage impedance of the transformers needs to be included in fault studies. On a per-unit basis, the transformer turns ratio normally does not appear in calculations. Transformer leakage reactances are the same for the positive and negative sequences. The zero-sequence impedance depends on how the three-phase windings are connected. For example, a delta-connected winding does not provide a path to the zero-sequence currents, as shown in Figure 13.8a.

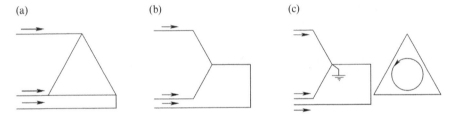

FIGURE 13.8   Path for zero-sequence currents in transformers.

Similarly, the Y-connected windings with an isolated neutral will appear as open-circuit to zero sequences, as shown in Figure 13.8b. As shown in Figure 13.8c, a grounded Y with delta-connected secondary windings provides a short-circuit path that allows zero-sequence currents to flow. This will also be the case if the secondary is connected in a grounded Y. However, the grounded-Y primary in Figure 13.8c will appear as an open circuit if the secondary is Y-connected with an isolated neutral. If zero-sequence currents can flow, a neutral-to-ground impedance Z_n, as shown in Figure 13.9a, will appear as $3Z_n$ in the zero sequence network in Figure 13.9b.

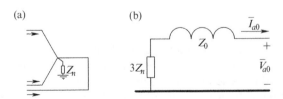

FIGURE 13.9   (a) Neutral grounded through an impedance; (b) zero sequence network.

In the case of three single-phase transformers, the zero-sequence impedance Z_0 in Figure 13.9b equals the positive-sequence leakage impedance. This is also the case in three-phase transformers with shell-type construction.

Example 13.2

Consider a simple system with a 1 pu load (where the load neutral is grounded) at bus 3 being supplied by a single generator, as shown in Figure 13.10, where all the quantities are in per-unit.

Calculate the fault-current magnitude in per-unit at bus 2 for (a) a three-phase fault and (b) a single-line-to-ground (SLG) fault, with the fault impedance as zero. Note that the per-unit values are on a common MVA base, and the reactances are in per-unit base impedances calculated at the low and the high voltage levels corresponding to the delta and Y sides of the transformer, respectively.

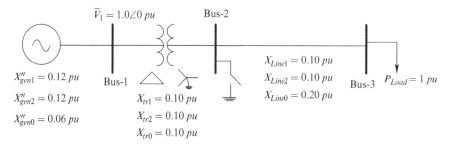

$\bar{V}_1 = 1.0\angle 0\ pu$ Bus-2

$X''_{gen1} = 0.12\ pu$ Bus-1

$X''_{gen2} = 0.12\ pu$ $X_{tr1} = 0.10\ pu$

$X''_{gen0} = 0.06\ pu$ $X_{tr2} = 0.10\ pu$

$X_{tr0} = 0.10\ pu$

$X_{Line1} = 0.10\ pu$

$X_{Line2} = 0.10\ pu$ Bus-3 $P_{Load} = 1\ pu$

$X_{Line0} = 0.20\ pu$

FIGURE 13.10 One-line diagram of a simple power system.

Solution MATLAB and Python solution is provided on the accompanying website. Consider the bus 1 voltage as a reference with $\bar{V}_1 = 1.0\angle 0\,\text{pu}$. The positive-sequence per-phase circuit is shown in Figure 13.11. Before the fault, $\bar{E}$ at the back of the subtransient reactance can be calculated by first calculating $\bar{I}_{\text{Load}}$.

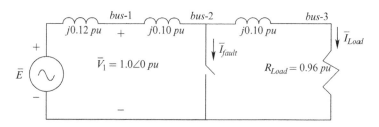

bus-1   bus-2   bus-3

$j0.12\ pu$ + $j0.10\ pu$ $j0.10\ pu$ $\bar{I}_{Load}$

$\bar{V}_1 = 1.0\angle 0\ pu$ $\bar{I}_{fault}$

$\bar{E}$ $R_{Load} = 0.96\ pu$

FIGURE 13.11 Positive-sequence circuit for calculating a three-phase fault on bus 2.

a) A three-phase-to-ground fault on bus 2 can be represented by closing the switch in Figure 13.11. Thus, the fault-current magnitude can be calculated as $I_{\text{fault}} = 4.69\angle -83.33$ degrees pu.

b) In the case of a SLG fault on bus 2, the sequence per-phase networks are
    connected in series, as the analysis in Figure 13.5 shows. With the fault
    impedance $Z_f = 0$, the circuit diagram is shown in Figure 13.12. In the
    zero-sequence network, the generator's zero-sequence impedance is shorted
    out because of the grounded-Y-delta connection. From Figure 13.12, the
    fault-current magnitude is $I_{\text{fault}} = 5.71\angle -84.63$ degrees pu.

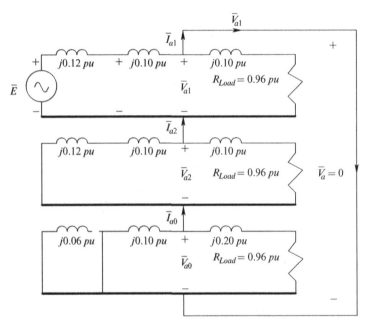

FIGURE 13.12 Sequence networks for calculating the fault current due to the
SLG fault on bus 2.

13.5   CALCULATING FAULT CURRENTS IN
LARGE NETWORKS

Example 13.2 shows the underlying principles for fault calculations that can be
used to analyze a power system with a few buses. In most practical networks
with thousands of buses, computer programs have been developed to carry out
these calculations. One of the procedures used in computing fault currents in
such a network is described in this section [3, 4].

Earlier, we discussed the nodal-equation formulation of the network for
the positive sequence

$$\overline{I}_{pos} = Y_{pos}\overline{V}_{pos} \tag{13.27}$$

where the subscript *pos* indicates a positive sequence in place of 1, which may be a bus number. Equation 13.27 can be written in terms of the impedance matrix as

$$\bar{V}_{pos} = Z_{pos} \bar{I}_{pos} \qquad (13.28)$$

where $Z_{pos} (= Y_{pos}^{-1})$ is the inverse of the Y-matrix. The impedance matrix in large networks is created without taking the inverse of the Y-matrix. Similar equations can be written to relate the negative-sequence and zero-sequence voltages and currents. From these networks, the fault currents for any type of fault can be computed using a computer program such as PSS®E [4].

Example 13.3
Consider the example three-bus power system discussed in earlier chapters and repeated in Figure 13.13. The modeling of this system under the pre-fault operating conditions is described on the accompanying website. A SLG fault occurs on line 1–2, one-third of the distance from bus 1. Calculate the fault current and various line currents.

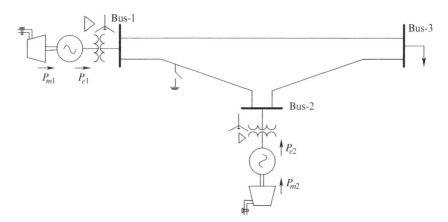

FIGURE 13.13 A SLG fault in the example three-bus power system for Example 13.3.

Solution The solution to this example is calculated by a MATLAB and Python program and is included on the accompanying website.

13.6 PROTECTION AGAINST SHORT-CIRCUIT FAULTS

Although open circuits sometimes can result in hazardous safety situations for personnel and must be guarded against, the overcurrent phenomena due to short-circuit faults is our focus here. To minimize the interval of the power disturbance and, more importantly, prevent power equipment from permanent damage, it is important that fault currents larger than the load currents for which the equipment is designed not be allowed to flow for intervals that can be destructive. For this purpose, all power system equipment is equipped with overcurrent protection [5–7]. The entire power system is divided into overlapping zones such that no part of it is left unprotected. This protection equipment can be categorized as follows, as shown in the block diagram in Figure 13.14 for one of the circuit breakers (CB).

- Current and voltage transformers (CTs and PTs) for sensing power-system voltages and currents
- Relays (Rs) that determine if a fault has occurred and issue a command to circuit breakers to operate
- Circuit breakers (CBs) that open the circuit contacts to interrupt the fault current and subsequently reclose them to resume normal operation

All of these are described briefly in the following subsections.

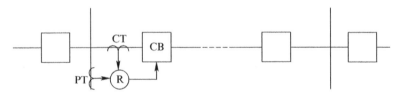

FIGURE 13.14   Protection equipment.

13.6.1   Current and Voltage Transformers

Currents and voltages in a power system must be sensed so that relays can determine if a fault has occurred. However, these voltages and currents are at extremely high values and must be stepped down to low-voltage signals with reference to a logic ground.

The CT, as shown in Figure 13.15, is a transformer in which the power system current flows through its primary, which usually has a single turn. The

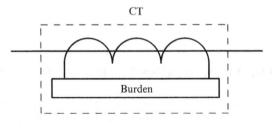

FIGURE 13.15   Current transformer (CT).

secondary generally has a large number of turns and produces a much smaller current that equals the primary current divided by the turns ratio, which flows through a small load referred to as *burden*. The burden is the relay connected to the CT secondary.

A capacitor-coupled voltage transformer (CCVT), one of several types, is shown in Figure 13.16. It uses the capacitive voltage-divider principle, where the output voltage of this divider is isolated through a transformer for safety purposes.

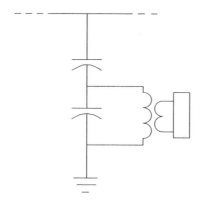

FIGURE 13.16   Capacitor-coupled voltage transformer (CCVT).

13.6.2   Relays

Based on sensed voltages and currents and other signals, relays decide if a fault has occurred and whether it should be interrupted by the circuit breaker. It is important that relays operate when they should, to protect the power system; but it is equally important that they not operate falsely, to avoid causing unnecessary power disturbances. Therefore, in relays, three things are important: selectivity, speed, and reliability.

13.6.2.1   Relay Types
There are many types of relays, but they can be basically categorized in the following manner.

Differential Relays. These relays may be used, for example, to protect a generator, bus, or transformer against internal faults, as shown in Figure 13.17 for protecting a bus.

Under normal conditions, as shown in Figure 13.17, the differential current through the relay, which is the difference between the measured currents, is zero. This is not so under an internal fault condition, causing the fault current to trip the circuit breaker.

Overcurrent Relays. In these relays, if the current being measured exceeds a minimum value greater than the maximum load current by a certain factor,

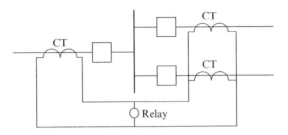

FIGURE 13.17 Differential relay for bus protection.

the relay determines that a fault has occurred and gives a "trip" command to the circuit breaker to operate. Generally, such relays operate on a delay-time basis. The delay time is an inverse function of the magnitude of the fault current: the larger the current magnitude, the shorter the delay time, as shown in Figure 13.18. It is possible to have several settings (1, 2, 3, and so on), as shown in Figure 13.18; thus, for the same current, the relay can operate with different time delays. These settings allow such relays to be coordinated with other relays protecting the system.

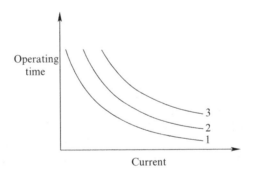

FIGURE 13.18 Time-current characteristics of overcurrent relays.

Directional Overcurrent Relays. The protection offered by these relays is for faults only in one direction. For example, in Figure 13.19, if the fault occurs on the right side of the CT location, the current $\bar{I}$ sensed by this relay is within +/− 90 degrees with respect to the voltage at this bus, causing the relay to trip

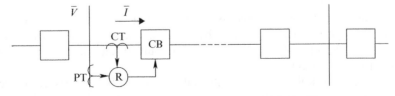

FIGURE 13.19 Directional overcurrent relay.

the circuit breaker. But for the fault on the left of the CT location, the current reverses from the previous case with current greater than +/− 90 degrees with respect to the voltage, and the relay will be blocked from tripping the circuit breaker.

Ground Directional Overcurrent Relays. These are zero-sequence relays that, as Figure 13.20 shows for the relay at bus A, act instantaneously by issuing a "trip" command to the circuit breaker if the fault is within 80% of line section A–B (considered its zone 1 of protection, discussed later); otherwise, the tripping time is increased as shown. The overcurrent relay at bus A is coordinated (time-delayed) with adjacent line-section relays. This ensures, for example, that the relay at bus A does not trip for faults in the adjacent line section B–C; it trips only as a backup to the relay at bus B, looking into line section B–C, if that relay malfunctions and fails to trip. Note that for a fault on the remaining 20% of line section A–B, the relay at bus A is also time-delayed, as shown in Figure 13.20; for such a fault, the relay at bus B, looking toward bus A, gives an instantaneous "trip" command to its circuit breaker because this fault is in its zone 1 of protection.

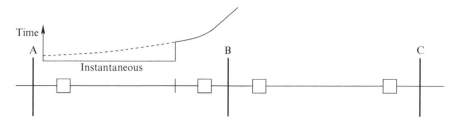

FIGURE 13.20   Ground directional overcurrent relay.

Directional Distance (Impedance) Relays. By calculating the ratio of the measured voltage and current, as shown in Figure 13.19, these directional relays determine the impedance and hence are called impedance relays. The characteristic of such relays is plotted in an $R-X$ plane, as shown in Figure 13.21, and goes through the origin. If the calculated impedance falls within the circle, the relay trips; otherwise, it is blocked.

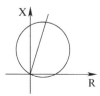

FIGURE 13.21   Directional impedance (distance) relay with an mho characteristic.

Under normal system conditions, load currents are much smaller than fault currents; hence the impedance (ratio of the measured voltage and current) is much larger and outside the circle in Figure 13.21, and the relay is blocked from tripping. In case of a fault – for example, on the right of the CT location on the line in Figure 13.19 – the measured impedance is the impedance of the line between the relay location and the fault; it is small – for example, somewhere along the straight line in Figure 13.21 – and within the circle, causing the relay to trip. This impedance also indicates the distance along the line where the fault has occurred.

The directionality of this relay is given by shifting the characteristic such that it passes through the origin, as shown in Figure 13.21. If the fault occurs to the left of the CT in Figure 13.19, for example, the impedance based on the measured voltage and current is in the third quadrant of the $R - X$ plane, outside the circle, implying that the relay will be blocked from tripping. Such relays are referred to as having an *mho* characteristic.

Pilot Relays. Pilot relays use a communication channel, a power-line carrier, or fiber optics to communicate between the two terminals of the transmission line being protected. If a fault occurs that is internal to the transmission line, the relays issue commands to circuit breakers at both ends of the line to interrupt the fault.

13.6.2.2   Zones of Protection in Transmission Lines

The entire power system is divided into protection zones: for example, as shown in Figure 13.22 for the relay at bus A.

Each zone encompasses one or more pieces of power-system equipment, and adjacent zones overlap with other relays so that no part of the power system is left unprotected even if one or more relays fail to operate:

- Zone 1: The first zone for the relay at A encompasses 80% of line A–B, for example. If the fault occurs in A's first zone, the relay acts instantly without any time delay. The remaining 20% of line A–B is protected by another relay or relay element at bus A but time-delayed for tripping.

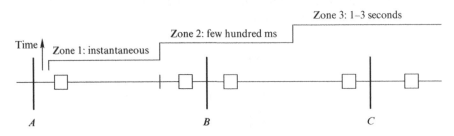

FIGURE 13.22   Zones of protection for the relay at bus A.

- Zone 2: The second zone for the relay at A encompasses the remaining 20% of line section A–B and overreaches in the next section B–C but below 80% of the zone 1 reach of the relay at B on line section B–C. If the fault occurs in the second zone assigned to this relay at bus A, it operates with a time delay of a few hundred milliseconds to prevent misoperation. Therefore, this relay must be coordinated with the relay at bus B, looking toward bus C, for which 80% of line B–C is its first zone.
- Zone 3: This adjacent zone in this relay at A is set to provide backup protection beyond zone 2 for the rest of line B–C and into the next line, with a time delay of 1–3 seconds.

13.6.2.3 Protection of Generators and Transformers

Similar to bus protection using a differential relay, shown in Figure 13.17, the differential relay shown in Figure 13.23 can protect a generator. A similar scheme can be used to protect transformers.

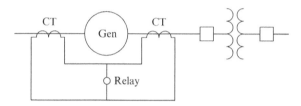

FIGURE 13.23 Protecting a generator using a differential relay.

Example 13.4

Consider the example three-bus power system shown in Figure 13.24. What types of relays can protect the generator and transformer at bus A? What types of relays act on the circuit breaker at bus A, looking toward bus B?

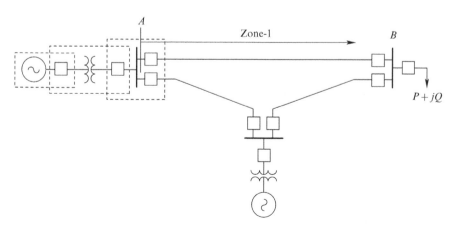

FIGURE 13.24 Relay for the three-bus power system in Example 13.4.

Solution The differential protection relay scheme shown in Figures 13.17 and 13.23 can protect the generator, the step-up transformer, and the bus. The circuit breaker at bus A on line A–B is acted upon by several relays: a differential relay (Figure 13.17) to protect the bus; an impedance relay (Figure 13.20) to protect against phase-phase faults (not involving ground) between phases *a-b*, *b-c*, and *c-a*; and overcurrent relays for SLG faults. Only protection zone 1 is shown in Figure 13.24.

13.6.3 Circuit Breakers

Circuit breakers are large pieces of equipment that, commanded by relays, interrupt the flow of current and break the circuit, as the name implies, to protect power equipment from short-circuit faults. There are circuit breakers available that can operate in two cycles, and further improvements are continuously being made. Various principles are used to elongate and cool the arc established by the parting contacts as they try to interrupt the current through them; this process is helped in the case of AC circuits, where the current naturally goes to zero every half-cycle.

Circuit breakers use a method of arc interruption based on their voltage level. At 345 kV and above, most circuit breakers use sulfur hexafluoride (SF_6) and a SF_6 gas puffer. SF_6 is a greenhouse gas, but there is no good substitute.

13.6.3.1 Automatic Reclosure

High-speed reclosing is important for stability. Since most faults are transitory, many utilities allow extra-high voltage (EHV) circuit breakers a single reclosing automatically. If the fault persists, the circuit breakers trip the line without further attempts to reclose automatically, and only the system operator can reclose the line. Such automatic reclosing is not recommended for lines leaving a generating station since reclosing into a persistent fault can fatigue the shaft of the turbine generator. In distribution systems, where it is important to maintain continuity of service to customer loads, several automatic reclosing attempts may be allowed.

13.6.3.2 Single-Phase (Independent-Pole) Operation

In conventional relaying and circuit-breaker schemes, all three phases are tripped for any type of fault. However, most faults involve only one of the phases, whereas tripping all three phases represents a more serious interruption. Therefore, certain utilities opt to open only the faulted phase at EHV and ultra-high voltage (UHV) levels. Of course, these schemes are more complex and costly than conventional approaches.

13.6.3.3 Circuit-Breaker Ratings

Circuit breakers, as discussed, are rated for their interrupting times in cycles of the line frequency. They have voltage ratings based on their insulation and current ratings based on the current they can safely carry and interrupt.

13.6.3.4 Symmetrical and Asymmetrical Current Ratings

A sudden fault in a power system can result in current transients that decay based on the X/R ratio of the circuit reactance X to the resistance R. The fault calculations discussed earlier in the chapter determined the *symmetrical* (with equal positive and negative peak values) line-frequency current following a fault. However, with fast circuit breakers, it is important to take into account the initial current offset that results in *asymmetrical* current, so-called because the positive and negative peak values are not the same. Asymmetrical current can be higher than symmetrical current by a factor depending on the reactance over the resistance ratio, X/R, of the network.

This initial current offset that decays with time can be understood by a simple R–L circuit with a sinusoidal source, as shown in Figure 13.25a, where $v_s(t) = \hat{V} \sin(\omega t + \beta)$.

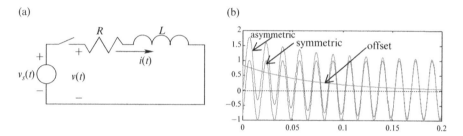

FIGURE 13.25 Current in an $R-L$ circuit. (a) Closing of the switch; (b) resulting current.

Choosing the switch-closing time as $t = 0$, the voltage $v(t)$ across the impedance can be expressed as $v(t) = \hat{V} \sin(\omega t + \beta)$. Therefore, in the circuit in Figure 13.25a,

$$Ri + L\frac{di}{dt} = \hat{V}\sin(\omega t + \beta) \quad (t > 0) \tag{13.29}$$

The natural current component, recognizing that this component decays with time, is

$$i_n(t) = Ae^{-t/\tau} \quad \text{(where } \tau = L/R \text{ is the time constant)} \tag{13.30}$$

The steady-state AC current component in this $R-L$ circuit is

$$i_{ac} = \hat{I}_{ac}\sin(\omega t + \beta - \phi) \tag{13.31}$$

where $\hat{I}_{ac} = \dfrac{\hat{V}}{\sqrt{R^2 + (\omega L)^2}}$ and $\phi = \tan^{-1}\left(\dfrac{\omega L}{R}\right)$.

Therefore, the complete current solution, using Equations 13.30 and 13.31, is

$$i(t) = Ae^{-t/\tau} + \hat{I}_{ac}\sin(\omega t + \beta - \phi) \tag{13.32}$$

where the unknown A can be calculated from the initial condition: that is, at time $t = 0$, the current $i = 0$. Therefore, from Equation 13.32, $A = \hat{I}_{ac}\sin(\phi - \beta)$, and thus

$$i(t) = \underbrace{\hat{I}_{ac}\sin(\phi - \beta)e^{-t/\tau}}_{\text{offset } I_{dc}(t)} + \hat{I}_{ac}\sin(\omega t + \beta - \phi) \quad (t > 0) \tag{13.33}$$

This asymmetrical current $i(t)$, where the decaying DC offset is superimposed on the AC (symmetrical) component, is plotted in Figure 13.25b. As can be seen from Equation 13.33, the offset $I_{dc}(t)$ is a DC component that decays exponentially with time as $e^{-(R/L)t}$, where the circuit time constant $\tau = L/R$. In low-resistance networks typical of power systems, the DC component decays slowly compared to the power frequency and can be assumed constant at the time of the circuit-breaker opening at a value I_{dc} over one or two power-frequency cycles. Therefore, the rms value I_{rms} of the asymmetrical current can be calculated, recognizing that the rms value of the AC component is $I_{ac} = \hat{I}_{ac}/\sqrt{2}$, as

$$I_{rms} = \sqrt{I_{dc}^2 + I_{ac}^2} \tag{13.34}$$

This discussion shows that in low-resistance networks that are typical of power systems, high-speed circuit breakers that interrupt current soon after a fault have to interrupt a current that is factor S higher than the AC current, where $S = 1.2$ for a two-cycle breaker (this includes a half-cycle minimum relay time plus the opening time of the breaker contacts) [8]. This factor S varies inversely, proportional to the breaker contact parting time, as defined in the ANSI/IEEE Standard C37.010 Application Guide [8]. For circuit breakers at above 115 kV voltage levels, the closing and latching capabilities for momentary symmetrical rms current are defined to be 1.6 times the rated rms short-circuit current [9].

REFERENCES

1. C. L. Fortescue. 1918. "Method of Symmetrical Coordinates Applied to the Solution of Polyphase Networks." *AIEE*. Vol. 37, 1027–1140.
2. P. Kundur. 1994. *Power System Stability and Control*. McGraw-Hill.
3. P. Anderson. 1995. *Analysis of Faulted Power Systems*. IEEE Press.
4. PSS®E. https://new.siemens.com/global/en/products/energy/energy-automation-and-smart-grid/pss-software/pss-e.html.
5. U.S. Department of Agriculture, Rural Utilities Service. *Design Guide for Rural Substations*. RUS Bulletin 1724E-300.
6. H. M. Rustebakke (ed.) 1983. *Electric Utility Systems and Practices*, 4th ed. John Wiley & Sons.

7. A. Phadke and J. Thorp. 2005. *Computer Relaying for Power Systems*. Institute of Physics Publishers.
8. H. O. Simmons Jr. "Symmetrical versus Total Current Rating of Power Circuit Breakers." *Applications of Power Circuit Breakers*. IEEE Tutorial Course, 75CH0975-3-PWR.
9. ANSI/IEEE. Standard C37.04.

PROBLEMS

13.1 Due to a single-line-to-ground fault on phase a in a system, $\overline{I}_{fa} = 5.0\angle 0$ pu and $\overline{I}_{fb} = \overline{I}_{fc} = 0$. Calculate the symmetrical components $\overline{I}_{fa1}$, $\overline{I}_{fa2}$, and $\overline{I}_{fa0}$ of the fault current.

13.2 Due to a line-to-line fault between phase b and phase c, $\overline{I}_{fa} = 0$ and $\overline{I}_{fb} = -\overline{I}_{fc} = 5.0\angle 0$ pu. Calculate the symmetrical components $\overline{I}_{fa1}$, $\overline{I}_{fa2}$, and $\overline{I}_{fa0}$.

13.3 At a point f in a system, there is an open-circuit fault on phase a with a voltage $\overline{V}_{fa}$ across the open circuit, whereas $\overline{V}_{fb} = \overline{V}_{fc} = 0$. Similar to calculating short-circuit currents using the sequence networks, calculate the sequence components $\overline{V}_{fa1}$, $\overline{V}_{fa2}$, and $\overline{V}_{fa0}$ at the fault point.

13.4 Repeat Example 13.2 if the single-line-to-ground fault is through a fault impedance $Z_f = 0.15\angle 0$ pu.

13.5 Repeat Example 13.2 if, before the fault, the load is zero: that is, $P_{\text{Load}} = 0$.

13.6 Repeat Example 13.2 if it is a double-line fault between phases b and c.

13.7 Repeat Example 13.2 if it is a double-line fault between phases b and c with a fault impedance $Z_f = 0.15\angle 0$ pu.

13.8 Repeat Example 13.2 if it is a double-line-to-ground fault with phases b and c grounded.

13.9 Repeat Example 13.2 if it is a double-line-to-ground fault with phases b and c grounded through a fault impedance $Z_f = 0.15\angle 0$ pu.

13.10 Repeat Example 13.2 if there is a single-line-to-ground fault on phase a of bus 1. The generator neutral is grounded through a resistance $R_n = 0.10$ pu.

13.11 Repeat Example 13.2 if there is a line-line-to-ground fault involving phases b and c of bus 1. The generator neutral is grounded through a resistance $R_n = 0.10$ pu.

13.12 Repeat Example 13.2 for an open-circuit fault by calculating the sequence voltages at the fault point on line 2–3 near bus 2, where the contact opens on phase a.

13.13 Given an impedance relay with the characteristic shown in Figure 13.21 applied to protect line 2–3 in Example 13.2, calculate the point in the impedance plane for a three-phase fault that is 85% of the line length away from bus 2. Repeat this if the fault is 15% of the line length away from bus 2.

13.14 Prove Equation 13.34.

14

TRANSIENT OVERVOLTAGES, SURGE PROTECTION, AND INSULATION COORDINATION

14.1 INTRODUCTION

Transmission and distribution lines form a large network spread over thousands of miles. A variety of reasons discussed in this chapter can cause abnormally high overvoltages that, unless properly protected against, can disrupt service momentarily at best and cause prolonged outages and expensive damage to power-system equipment at worst. In this chapter, we will examine the causes of overvoltages and the measures that can be taken to protect against them.

14.2 CAUSES OF OVERVOLTAGES

Overvoltages are caused primarily by lightning strikes and switching of extra-high voltage transmission lines, both of which are examined in the following subsections.

Electric Power Systems with Renewables: Simulations Using PSS®E, Second Edition. Ned Mohan and Swaroop Guggilam.

14.2.1 Lightning Strikes

As mentioned earlier, transmission lines, stretched over long distances, are frequently subjected to lightning strikes. The frequency of such strikes depends on their geographical location. Lightning is not a very well understood phenomenon, but it suffices here to say that a lightning strike to equipment results in a brief discharge of current pulse to it with respect to ground. This pulse is commonly said to reach its peak at time t_1 and exponentially tapers off to half its peak value at t_2. Peak current strikes as high as 200 kA have been recorded, although generally, a peak of 10 to 20 kA is assumed to protect against the resulting overvoltages.

14.2.1.1 Lightning Strikes to Shield Wires

Many transmission lines have shield wires, as shown in the tower structure in Figure 14.1a. These shield wires are located higher than the transmission-line conductors and hence protect them from being hit directly by lightning strikes. They provide an approximately 30-degree protection zone around them. Shield wires are grounded through the tower, and it is desirable to keep the tower footing impedance as small as possible. However, not all transmission systems have shield wires; many utilities find it more economical to use surge arresters more extensively (although not at every tower) rather than employ shield wires.

(a)

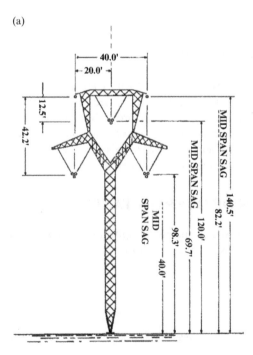

FIGURE 14.1   Lightning strike to the shield wire. (a) Tower configuration; (b) lightning strike.

(Continued)

(b)

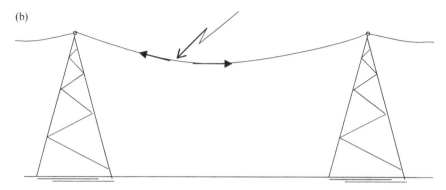

FIGURE 14.1 (Cont'd)

When lightning strikes a shield wire, the resulting current waves flow in both directions and pass through the towers to ground, as shown in Figure 14.1b. The tower footing resistance and the $L(di / dt)$ effect due to the rapidly rising current front may cause the tower potential to exceed the insulation strength of the insulator string, causing it to flash over (called *backflash*).

14.2.1.2 Lightning Strike to a Conductor

Another scenario is where a conductor is struck by lightning. The resulting traveling waves proceed in both directions, and insulator strings may flash over; otherwise, when the current wave reaches its termination, surge arresters (discussed later) prevent the voltages from rising to a level that can damage the equipment [2].

14.2.2 Switching Surges

At extra-high voltage levels and above, switching of transmission lines, as shown in Figure 14.2a, can result in overvoltages (Figure 14.2b) that can be higher than those caused by lightning strikes. We will see later that the

(a)

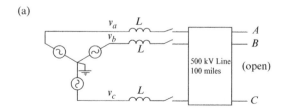

FIGURE 14.2 Overvoltages in per-unit due to switching of transmission lines. (a) Circuit diagram; (b) switching voltage waveforms.

(Continued)

(b)

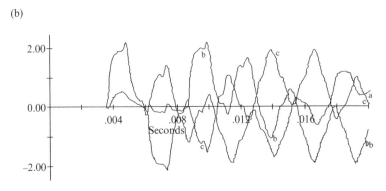

FIGURE 14.2   (Cont'd)

insulation requirement is influenced by both the amplitude and the duration of the overvoltages. These switching overvoltages can be minimized by using pre-insertion resistors, as shown in Figure 14.2b: before switching in a transmission line, a resistor is inserted in series; it is bypassed later for normal operation.

14.3   TRANSMISSION-LINE CHARACTERISTICS AND REPRESENTATION

The discussion of transmission lines here also includes distribution lines. In the following analysis, we will assume a transposed three-phase line and utilize the discussion in Chapter 4 dealing with transmission lines in a steady state. Assuming this line to be lossless, the voltage and current magnitudes of traveling waves are related by the characteristic impedance Z_c of this transmission line, where, as calculated in Chapter 4,

$$Z_c = \sqrt{\frac{L}{C}} \tag{14.1}$$

and L and C are the inductance and capacitance per-unit length of the transposed three-phase lines [3]. These traveling waves propagate at a velocity c that is related to the transmission-line parameters in the following manner:

$$c = \sqrt{\frac{1}{LC}} \tag{14.2}$$

From Chapter 4, the per-phase inductance per-unit length is

$$L = 2 \times 10^{-7} \ln \frac{\sqrt[3]{D_{12} D_{23} D_{31}}}{r} \ \text{H/m} \tag{14.3}$$

where the current is assumed to be at the outer surface of the conductor with a radius r.

Also from Chapter 4, the per-phase capacitance per-unit length in air is

$$C = \frac{2\pi \times 8.85 \times 10^{-12}}{\ln \frac{\sqrt[3]{D_{12}D_{23}D_{31}}}{r}} \text{ F/m} \qquad (14.4)$$

Using Equations 14.3 and 14.4, the traveling wave speed in Equation 14.2 is $c = 3 \times 10^8$ m/s, which is the speed of light. This speed is less for practical transmission lines with losses, particularly in zero-sequence mode.

Transient disturbances usually are not balanced; for example, only one of the transmission-line phases is normally subjected to a lightning impulse. Therefore, the zero-sequence path involving ground return must be carefully included in modeling to get correct overvoltages. Similarly, transmission lines are not perfectly balanced, and most lines are not transposed, so they may need to be represented as un-transposed. In addition, we are no longer dealing with a steady-state line frequency of 60 or 50 Hz but rather transient phenomena involving very high frequencies. Figure 14.3 illustrates the frequency dependence of the line parameters for positive- and negative-sequence quantities and zero-sequence quantities. This frequency dependence should be included for an accurate analysis.

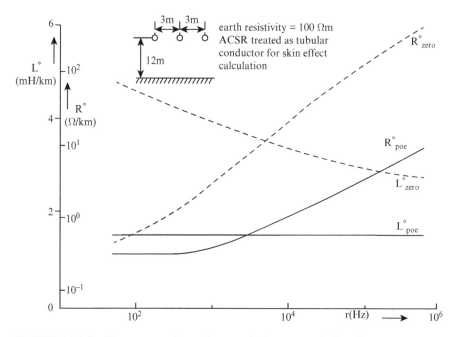

FIGURE 14.3 Frequency dependence of the transmission-line parameters. Source: [1].

14.3.1 Calculating Overvoltages

In very simple cases, overvoltages can be calculated by keeping track of traveling waves using a *Bewley diagram* (named after L. V. Bewley of the General Electric Company, who pioneered it) or with Laplace transforms. However, even the simplest three-phase system without transposition and frequency dependence of parameters is too complex to be amenable to such methods, where traveling waves in various modes combine to result in the resultant voltage [1].

14.4 INSULATION TO WITHSTAND OVERVOLTAGES

Power-system equipment is designed with insulation to withstand a certain amount of voltage. The insulation level necessary to withstand voltages depends on factors such as the shape of the voltage wave and its duration and the condition of the insulation in terms of its age, humidity, and contamination levels. Power-system equipment can be categorized based on two types of insulations: self-restoring and non-self-restoring. Insulator strings on transmission lines represent the self-restoring type of insulation, which can be allowed to break down occasionally and which restores itself after the resulting fault clears. However, insulation for transformers, generators, and so on is the non-self-restoring type: if allowed to fail at any time, severe and permanent damage will occur that will be very expensive to repair.

For self-restoring insulation, such as the insulator strings of transmission lines, it is cost-prohibitive to insulate against all expected overvoltages, and hence a statistical approach may be used to determine an acceptable probability of failure. The non-self-restoring insulation in transformers is protected against failures using surge arresters, as discussed later in this chapter.

As mentioned earlier, the ability of insulation to withstand voltage depends on the shape of the voltage wave and its duration, which are different for lightning and switching surges. The voltage levels that given insulation can withstand are defined next.

14.4.1 Basic Insulation Level (BIL)

For a lightning impulse, the level of insulation is specified as the *basic insulation level* (BIL), which is the peak of the withstand voltage of a standard lightning impulse voltage wave. This lightning impulse wave, as shown in Figure 14.4, is assumed to rise to its peak value in $1.2\,\mu s$ and tapers off exponentially to one-half of its peak value in $50\,\mu s$.

As an example, the BIL for a 345 kV transformer is shown in Figure 14.5 as 1,175 kV, which is the peak line-to-ground voltage.

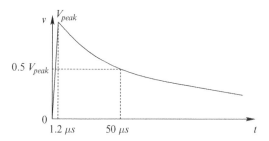

FIGURE 14.4   Standard voltage impulse wave to define the basic insulation level.

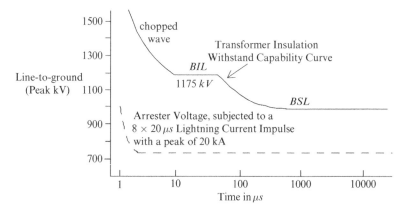

FIGURE 14.5   345 kV transformer voltage insulation levels.

14.4.2   Basic Switching Insulation Level (BSL)

As shown in Figure 14.2, voltages applied at one end of a transmission line result in traveling waves of voltage and current with steep fronts. These switching surges do not have as steep a front as those resulting from lightning impulses; however, switching surges are of longer duration. The standard switching impulse voltage wave is assumed to rise to a peak value in $250\,\mu s$ and tapers off exponentially to one-half of its peak value in $2,500\,\mu s$. The withstand insulation level for the peak of such a switching surge is called the *basic switching insulation level* (*BSL*). As shown in Figure 14.5, equipment with a particular BIL has a lower withstand capability to switching surges: that is, it has a lower BSL than BIL since switching surges have a longer duration.

14.4.3 Chopped-Wave Insulation Level

As shown in Figure 14.5, the same equipment insulation can withstand higher voltages than its BIL if the voltage impulse applied to the equipment is chopped to bring it to zero in 3–4 μs after the impulse has reached its peak value.

14.5 SURGE ARRESTERS AND INSULATION COORDINATION

In power-system equipment, insulation is protected against voltage impulses using surge arresters connected in parallel. The role of surge arresters is somewhat like that of Zener diodes in low-power electronic circuits. Surge arresters appear as open circuits at normal voltages and draw almost negligible current. However, by providing a low-impedance path, they allow whatever current needs to flow, of course within their rating by proper selection, thus "clamping" the voltage across the device being protected to a threshold voltage (plus a small *IR* current-discharge voltage), as shown in Figure 14.5 by the dotted curve. A surge arrester should be able to dissipate the associated energy following an impulse without any damage to itself, thus allowing a quick return to normal operation when it should once again appear as an open circuit. In power systems, modern practice is to use metal oxide (zinc oxide) arresters that have a highly nonlinear *i-v* characteristic such as that given by the following relationship:

$$i \propto V^q \tag{14.5}$$

where the voltage is expressed in per-unit of the overprotection voltage value and the exponent q can be as high as 26–30. Therefore, the current through the arrester for voltages substantially below unity is negligible. However, in the overvoltage protection region, the arrester voltage remains almost constant. The arrester discharge voltage level should be sufficiently below the equipment insulation level that the arrester is to protect and provide a margin given by various IEEE and ANSI standards. Therefore, insulation coordination requires coordinating and selecting BIL and BSL values for various equipment and the ratings of the arresters to protect them, as illustrated in Figure 14.5.

REFERENCES

1. H. W. Dommel. 1986. *EMTP Theory Book*. BPA, Contract No. DE-AC79-81BP31364.
2. *Surge Protection in Power Systems*. 1979. IEEE Tutorial Course, 79EH0144-6-PWR.
3. Electric Power Research Institute (EPRI). 1987. *Transmission Line Reference Book: 345 kV and Above*, 2nd ed.

PROBLEMS

14.1   What are the causes of overvoltages?

14.2   Why is it important to include the frequency dependence of the transmission-line parameters in calculating surge overvoltages?

14.3   What is meant by the basic insulation level?

14.4   What is meant by insulation coordination?

14.5   What is the purpose of surge arresters, and how do they work?

INDEX

3-phase, 28–29, 129

A
ACE (Area Control Error), 251–52, 254
AGC (automatic generation control), 247–48, 251–57
Area Control Error. *See* ACE
automatic generation control. *See* AGC
automatic voltage regulation. *See* AVR
average voltage, 160, 163–64
AVR (automatic voltage regulation), 208, 246–47

B
balanced three-phase circuits, 24–25, 28
base values, 32–33, 127–28
bus voltage magnitudes, 110, 216, 222

C
CB. *See* circuit breakers
change in load, 250–52, 262
circuit breakers (CB), 9, 174, 206, 230, 233, 235–36, 239, 265–87
coal-fired power plants, 54–56, 68
conductor radius, 77, 79, 90
conductors, 34, 72–81, 85, 197, 201, 291, 293
configurations, 27, 48
control areas, 247, 251–54, 260–61
Control of Interconnected Power Systems and Economic Dispatch, 247–63
converters, 62–63, 65, 68, 146–50, 152–57, 159, 161–62, 164, 166, 168–69, 171–72, 224–27
core losses, 124–25, 131, 141–42

CT (Current transformer), 278, 280, 282–83
currents, harmonic, 171, 190–91
Current transformer. *See* CT
current waveforms, 163–66, 169, 171, 183–84, 189, 192

D
DC-bus voltage, 157, 172, 224
DC-side voltage, 147, 164, 166, 169
degrees, 22–24, 26, 134, 155, 164, 166, 169, 171–72, 197–98, 201, 203, 205–6, 223, 237, 267
delta-connected load, balanced, 46–48
DERs (distributed energy resources), 5, 145
DG (distributed generation), 5, 9, 66, 173
Differential Relays, 279, 283–84
Displacement Power Factor. *See* DPF
distortion component, 183–87
distributed energy resources (DERs), 5, 145
distributed generation. *See* DG
Distribution System, Loads, 175–93
distribution systems, 4, 173, 284
double-line fault, 269, 271–72, 287
double-line-to-ground fault, 269, 271, 287
DPF (Displacement Power Factor), 188–89, 193
duration, 292, 294–95
DVRs (dynamic voltage restorer), 181, 192
dynamic voltage restorer. *See* DVRs
dynamic voltage restorers, 181, 192

E
EHV (extra-high voltage), 284, 289, 291
electricity, 1–2, 5–6, 8, 51, 53–54, 63, 65–66, 69, 71, 173, 175–76, 178–79
electricity generation, 2, 6, 51–52, 69

Electric Power Systems with Renewables: Simulations Using PSS®E, Second Edition. Ned Mohan and Swaroop Guggilam.
© 2023 John Wiley & Sons, Inc. Published 2023 by John Wiley & Sons, Inc.
Companion Website: www.wiley.com/go/mohaneps

Electric Power Research Institute (EPRI), 89–90, 296
EPRI (Electric Power Research Institute), 89–90, 296
equating, 27, 216, 261
Equation, 15–16, 18–19, 24–29, 31, 35–36, 38–44, 49–50, 75–87, 98–110, 112–15, 137–39, 149–50, 183–89, 203–4, 216–19, 230–35, 241–43, 250–52, 258–61, 270–73
equipment, power-system, 282, 289, 294, 296
extra-high voltage (EHV), 284, 289, 291

F
fault, 9, 174, 181, 208–10, 229–31, 233–37, 240–41, 265–66, 268–70, 273, 275–87
fault currents, 95, 266, 268, 273, 277–78, 282
faulted line, 231, 234–35
Faulted Power Systems, 141, 213, 286
faulted system, 267–69, 271–72
fault impedance Zf, 270, 276, 287
fault point, 268–69, 287
field intensity, 35–36, 47
FIGURE 2A, 47–48
FIGURE 7A, 158–60
FIGURE 7B, 162–65, 167–71
FIGURE 11A, 241–43
flux direction, 41–42
flux lines, 25, 37, 39, 43, 47, 196–97, 200, 209
Fourier Analysis, 184–86
frequency dependence, 74, 293–94, 297
fundamental frequency component, 184, 186–87, 191

G
generator models, 151, 211
generators
  equivalent, 151, 157, 254
  two-pole, 196–97
GFIs (ground fault interrupters), 174, 192
grid, 3, 5–6, 9, 61–63, 146–50, 201, 206–7, 211, 225, 247
ground, 80–81, 85, 98, 134, 161, 174, 265–66, 269–73, 290–91
ground fault interrupters (GFIs), 174, 192

H
HIGH-VOLTAGE DC, 4, 71, 145, 152
HVDC-VSC Systems, 145, 152, 225
hydro generators, 195–96

I
ideal transformer, 123–24, 126–27, 132, 141, 157–58, 161
impedance, zero-sequence, 273–74, 276
independent power producers (IPPs), 5, 9
induction generator, 61–63
input voltage, 29, 141, 163, 179
instantaneous power, 16–17, 28–29
Interconnected Power Systems, 6, 229, 239, 245–63
interconnected systems, 8–9, 180, 195, 229, 245, 249–50, 254, 257, 262
interface, power-electronics, 62, 146, 177–80
IPPs (independent power producers), 5, 9

K
kV line, 90–91, 219, 291
kV transformers, 129, 131, 294
kV transmission lines, 81, 85, 91–92, 97, 129, 233, 236, 238

L
LCCs (line-commutated converters), 8, 145, 152, 155, 161
leakage flux, 43–44, 204
leakage impedances, 124–25, 127–28, 131–32, 137, 274
line-commutated converters. *See* LCCs
line length, 86, 97, 287
line-to-line voltages, 26, 159
load, 19, 21, 23–24, 29–30, 33, 45–46, 123–24, 131–32, 141, 173–78, 181–83, 188, 190–93, 215–18, 220–21, 238–39, 246, 248–55, 259–60, 262–63
  electrical, 236, 248–50
  linear, 183, 188, 190
load change, 250–52, 254–55, 257, 261–62
load demand, 191, 247–48
load factor, 175, 192
load-frequency control, 248–49
load impedance, 21, 27, 29, 33, 91, 95
loading, 84–85, 89, 91, 220
load management, 191, 193
load tap changers (LTCs), 132, 181
long lines, 74, 86, 88, 91
LTCs (load tap changers), 132, 181

M
magnetic field, 34–36, 197, 202
magnitude, fault-current, 275–76
marginal costs, 258–62
MATLAB, 108, 112, 116, 233, 239, 255, 268, 275, 277

measured voltage and current, 281–82
medium-length lines, 86–87, 91–92
motor loads, 22, 34, 176–77
MW load, 253–54

O

OPF (optimal power flow), 9, 246, 261, 263
optimal power flow. *See* OPF
output voltage, 131–32, 141, 157, 159–61, 279
overvoltages, 289, 291–92, 294, 297

P

per-phase, 28–29, 85, 87–88, 134, 178, 204,
    209–10, 213
per-phase basis, 24, 33, 74, 82–83, 136, 148,
    183, 268, 270
per-phase circuit, 24, 29, 46
per-unit quantities, 11, 32–33
per-unit values, 32–33, 97–98, 130, 275
PF. *See* Power Factor
phase conductors, 76, 90
phase currents, 27, 46–47, 148, 171, 201,
    203–4, 268
phases, 16–18, 23–26, 28–29, 47–48, 72–74,
    76, 79–81, 157–59, 161, 165–66, 168,
    197–204, 226, 265–68, 270–71, 284, 287
phase shifters, 137–38
phase shifts, 8, 128, 134–35, 137–38
phase voltages, 24, 26–28, 45, 148, 161, 168,
    171, 226
pi-circuit, 138–39
PLoad, 248–50, 253, 256, 275, 287
post-fault, 231, 234–37
power, wind, 60, 63
power electronics, 8, 62–63, 68–69, 71, 89,
    157, 161, 179, 181, 191–92
power-electronics-based loads, 177, 179–80,
    182, 192
power electronics interface, 63,
    69, 146
power-electronics systems, 182–83, 189–91
power factor, 7, 11, 16, 19, 21, 28–29, 45–46,
    141, 168, 176–77, 182, 188–89, 220–21
Power Factor (PF), 16, 19, 21, 28–29, 45–46,
    62, 91, 132, 141–42, 168, 172, 176–77,
    182–83, 188–89, 193, 206, 219–21
power factor, unity, 21, 142, 177,
    180, 206
power flow, 8, 62, 95, 108–9, 112–13,
    130–31, 134, 140, 142–43, 150, 152,
    154–55, 172, 253–54, 261–62
power-flow calculations, 95–97, 113
power-flow Equations, 104, 111
Power Flow in Power System Networks,
    97–117

Power-flow results for Example, 109, 117,
    154
power level, 152, 257–58
power system loads, 174–75, 177, 181
power systems, 1–2, 4–9, 32–34, 96–97, 106,
    108–9, 111–12, 119–43, 215–46,
    252–53, 265–67, 278–79, 282, 285–87,
    296
    three-bus, 129, 228, 253, 277, 283
    three-phase, 22–23, 128, 134
Power System Stability and Control, 90, 112,
    141, 213, 227, 262, 286
power system stabilizer. *See* PSS
power triangle, 17–19, 29
problem, 45–47, 56, 69, 87, 89, 91–92, 96,
    113, 142, 172, 213, 228, 263
PSS (power system stabilizer), 221, 240, 247
PSS®E, 6–9, 92, 108–9, 112–13, 117, 130,
    140, 142–43, 145, 150–51, 153–54, 211,
    213, 215–17, 225, 228–29, 239–40,
    261–63
Python, 108, 112, 116, 233, 237, 239, 254,
    268, 275, 277

R

radial system, simple, 215–16, 219–20
reactive power, 16, 18–22, 28, 31, 62, 84,
    95–96, 105–6, 110, 113–14, 136,
    148–50, 152–53, 171–72, 206–8,
    215–18, 220–21, 225, 228, 245–46
receiving, 22, 30–31, 46, 82–83, 86–87,
    91–94, 155, 162, 215–18, 220
relays, 9, 236, 265–66, 278–84
    overcurrent, 279–81, 284
renewable energy, 53, 58, 146
residential loads, 173–75
RLoad, 275–76
rotor angle, 205, 230–31, 234–38, 241

S

SCR (short-circuit ratio), 162, 191
series impedance, 88, 98–99, 129–30
short-circuit faults, 208–9, 266, 269, 277, 284
short-circuit ratio. *See* SCR
SIL (surge impedance loading), 84–86, 89, 91,
    218, 220
Simulink, 257
single-line-to-ground fault, 269–70, 287
static var controllers. *See* SVCs
stator currents, 201–2, 209
Steam turbine, 195–96, 256
studies, power-flow, 129–30, 132, 138–39,
    142–43, 213–14
subcircuit, 16–17
Substituting Equation, 100, 184, 188

surge impedance, 83–85, 90–91, 93
surge impedance loading. *See* SIL
SVCs (static var controllers), 222–24, 246
synchronous, 60–62, 178–79, 195, 201, 203, 207–8, 230–32, 235, 242, 273
synchronous generators, 8, 34, 110, 119, 195–213, 220–21, 230, 240–41, 246, 273
system MVA base, 232–33, 237

T

TCR (Thyristor-controlled reactor), 223–24
TCSC (thyristor-controlled series capacitor), 225–26, 238, 246
THD (Total Harmonic Distortion), 183, 185–87, 189–93
three-bus system, 110, 113–14, 225, 238
three-phase circuits, 11, 22, 24, 26, 28–29
three-phase fault, 275, 287
three-phase system, 128–29, 268, 294
three-phase-to-ground faults, 269, 275
three-phase transformers, 136, 275
three-phase windings, 27, 197, 199, 274
Thyristor-controlled reactor. *See* TCR
thyristor-controlled series capacitor. *See* TCSC
thyristors, 154, 162–67, 169, 181, 222, 226
tie lines, 247, 249, 251–52
time delay, 280, 282–83
time domain, 13, 15, 23, 238
Total Harmonic Distortion. *See* THD
transformer loading, 131–32
transformers, 2, 8, 18–19, 22–23, 27, 32, 34, 96–97, 119–20, 122–25, 127–32, 134, 136–42, 170, 181–83, 213–14, 246, 273–75, 278–79, 283
   phase-shifting, 140, 143, 227
   single-phase, 136, 275
Transformers in Power Systems, 121–43
Transmission Line Faults, 265, 267–87
transmission-line parameters, 74, 80, 292–93, 297
transmission lines, faulted, 233, 235, 239
transmission system operators (TSOs), 5, 9
TSOs (transmission system operators), 5, 9
turbine, 8, 53–58, 60–62, 196, 204–7, 230, 241–42, 248, 250
two-area system, 255–57
two-winding transformers, 133, 136, 141–42

U

UHV (ultra-high voltage), 173, 284
ultra-high voltage (UHV), 173, 284
Unified Power-Flow Controller. *See* UPFC
uninterruptible power supplies (UPSs), 180, 192

UPFC (Unified Power-Flow Controller), 226–27
UPSs (uninterruptible power supplies), 180, 192
utility grid, 145–46, 204, 207

V

voltage and current, 12, 17–18, 127, 141, 156, 273, 295
voltage control, 132, 153, 221, 246
voltage equation, 83–84, 272
voltage levels, 73, 80, 85, 95, 119, 182, 246, 284, 294
voltage-link system, 177, 179–81
voltage phasor, 18, 207
voltage polarity, 11, 16, 41–42, 200–201
voltage profile, 84, 91, 218
Voltage Regulation and Stability in Power Systems, 217–27
voltages
   applied, 13, 15, 86, 120, 123–24, 141, 178
   induced, 40, 42, 124, 200–201
   line-line, 27, 90, 142, 164, 166
   measured, 281–82
   rated, 32–33, 122, 125, 127, 132, 173, 220–21
   receiving-end, 84, 215, 220
   terminal, 134, 213
   three-phase, 22–23, 25, 27, 157, 172
   withstand, 157, 294
voltages and currents, 11–12, 46, 127, 132, 142, 267–68, 278
voltage sensitivity, 176–77, 228
voltage source, 19, 21, 141, 157, 183, 190, 230, 274
voltage source converters. *See* VSCs
voltage-source converters, 145, 147, 151–52
voltage stability, 8, 215, 220, 226, 261
voltage supply, 180–81, 190
voltage transformation, 122, 137
voltage waveforms, 159, 165, 171, 180, 182, 190
VSCs (voltage source converters), 8, 141, 145, 147–50, 152–53, 157–59

W

waveforms, 17, 158, 164–69, 171, 174–75, 185–86
wind energy, 58, 68, 71
wind generators, 146, 150
winding, 25, 119–24, 131–33, 135–36, 141, 196–200, 202–4, 208–9, 246–47

Z

zero-sequence currents, 136, 274

Printed and bound by CPI Group (UK) Ltd, Croydon, CR0 4YY

16/04/2025

14658349-0001